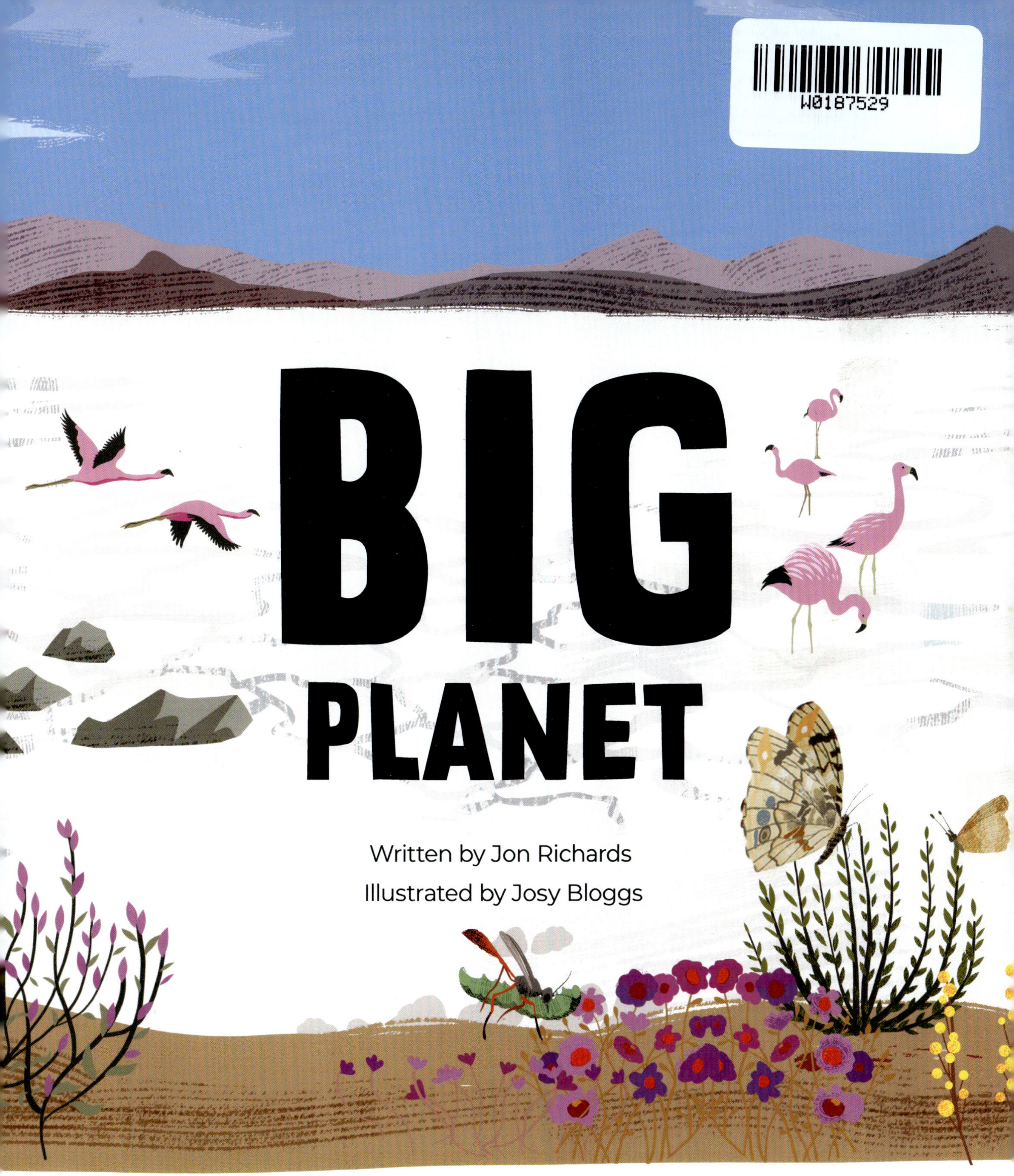

BIG
PLANET

Written by Jon Richards

Illustrated by Josy Bloggs

FRANKLIN WATTS
LONDON • SYDNEY

Natural
History
Museum

Contents

Big diverse planet

Earth is a planet of huge contrasts, from arid desert to dense forest and icy polar regions. These variations produce a broad range of habitats, each with their own amazing plants and animals.

Tropical zone
The equator is an imaginary line which runs around the middle of Earth. Either side of the equator are the tropics, where the weather is usually hot all year round. There is often a lot of rain in the tropics but there are also many dry regions.

Temperate zone
The temperate zones are found to the north and south of the tropics. Regions in this zone usually experience four seasons in a year – spring, summer, autumn and winter.

 **56.7°C** – the highest temperature ever recorded on Earth at Furnace Creek, California, USA

 **-89.2°C** – the lowest temperature ever recorded on Earth at Vostok Station, Antarctica

 Cherrapunji, India, is the place with the most rainfall in a year. During 1860–1861, **26,470 MM** of rain fell.

Polar zone

The polar zones lie at the northern and southern limits of the world. Conditions here can be very cold, especially during the long winter months.

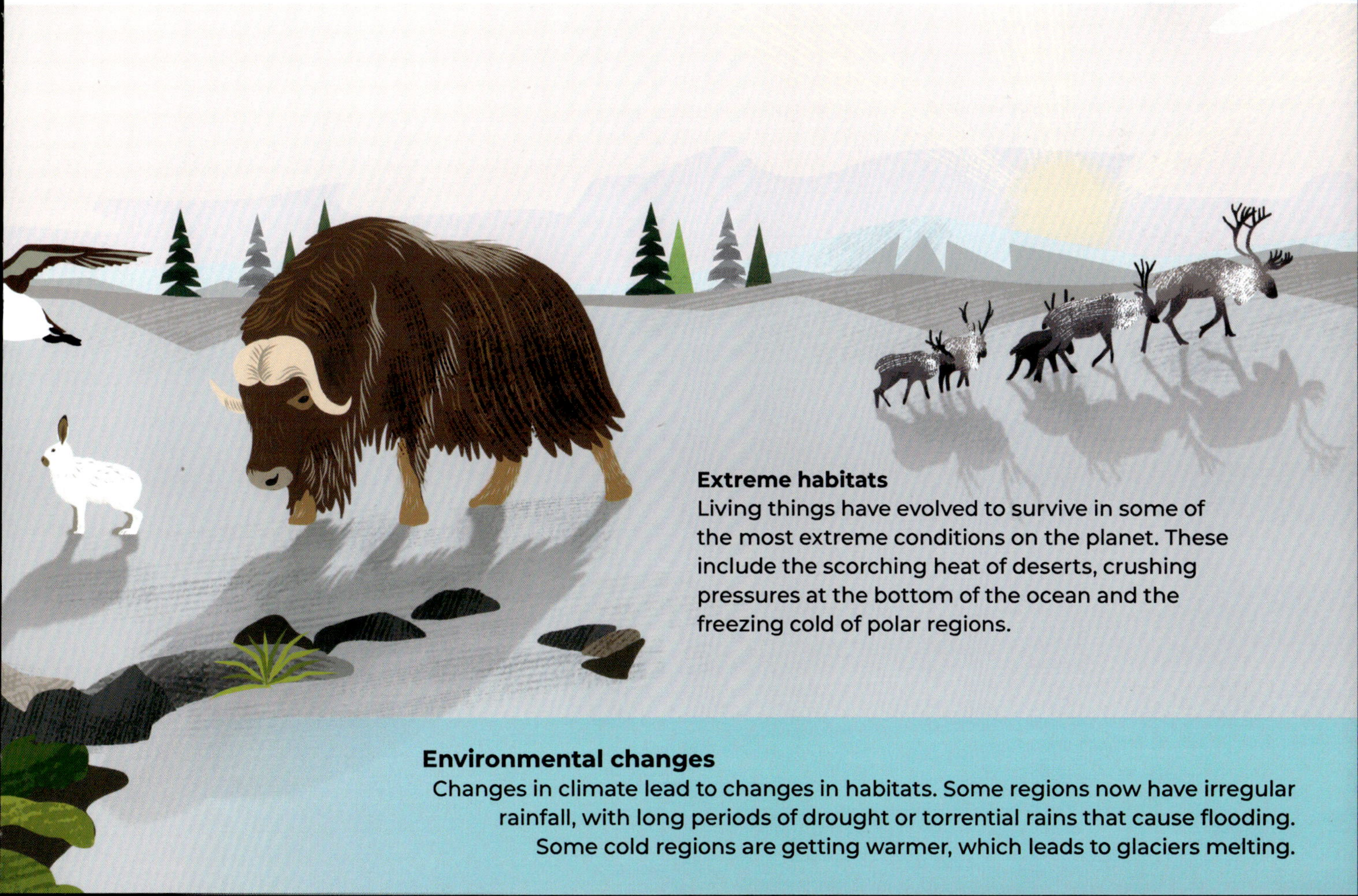

Extreme habitats

Living things have evolved to survive in some of the most extreme conditions on the planet. These include the scorching heat of deserts, crushing pressures at the bottom of the ocean and the freezing cold of polar regions.

Environmental changes

Changes in climate lead to changes in habitats. Some regions now have irregular rainfall, with long periods of drought or torrential rains that cause flooding. Some cold regions are getting warmer, which leads to glaciers melting.

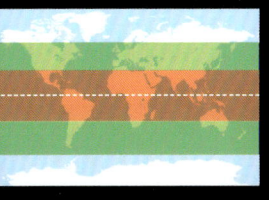

Polar zone
Temperate zone
Tropical zone
Temperate zone
Polar zone

29%

THE AMOUNT OF EARTH'S SURFACE THAT IS COVERED BY LAND

Of this, **31 per cent** is covered by desert (including the North and South Poles)...

... **33 per cent** is covered by grassland...

... **and 36 per cent** is covered by forest.

Tropical rainforest
THE AMAZON

Tropical rainforests are found in parts of the world that are warm and very wet. The biggest rainforest in the world is located in the Amazon River basin in South America.

Layered life
At the top of the rainforest are the tallest trees, which poke above the next layer, the canopy. Below the canopy is the understorey, made up of shrubs and small trees. At the bottom is the dark forest floor.

Forest plants
Many rainforest plants don't grow on the ground. These are known as epiphytes and they live on other plants. Rainforest trees include Brazil-nut trees and palms.

Flooding river
Every year, the Amazon River floods almost 250,000 square km of the forest. Tiny zooplankton are carried onto the flooded forest floor and attract fish and other predators.

Forest animals

There are millions of species of rainforest bugs and invertebrates, from tiny leaf-cutter ants to giant bird-eating spiders. Living in the branches are amphibians, such as poison dart frogs, and reptiles, such as boas. Mammals include giant otters and jaguars, while colourful macaws soar above the trees.

Dwindling forest

In the last 40 years, more than 18 per cent of the forest has been cleared to make room for homes and industry. Climate change has led to drier conditions, causing forest fires. This habitat loss threatens the survival of many species.

Amazon Rainforest

Amazon River

SOUTH AMERICA

6.7 MILLION SQUARE KM

THE AREA OF THE AMAZON RAINFOREST

- -

The Amazon accounts for about

HALF

of the world's tropical forest.

- -

THE AMAZON IS HOME TO ONE IN TEN SPECIES ON EARTH. THESE INCLUDE:

40,000 plant species

2.5 million insect species

3,000 species of freshwater fish

Nearly **1,300** bird species

370 species of reptile

More than **400** mammal species

Desert

ARID ATACAMA

Situated on the western edge of South America is one of the driest places on the planet. The Atacama Desert sits in the shadow of the towering Andes and stretches for about 1,000 km from north to south.

Rain shadow
The Atacama Desert is found between two chains of mountains – the Andes to the east and the Cordillera de la Costa to the west. Any wet air is forced up the mountains and the water it carries is dumped as rain on the mountainsides, leaving only dry air to reach the Atacama.

Dry and barren
Much of the desert floor is covered by stones, sand and salt pans – large areas of salt left behind after any water has evaporated.

SOUTH AMERICA

Atacama Desert

On average, the Atacama receives just 15 mm of rain a year.

SOME WEATHER STATIONS IN THE ATACAMA HAVE RECORDED NO RAINFALL AT ALL.

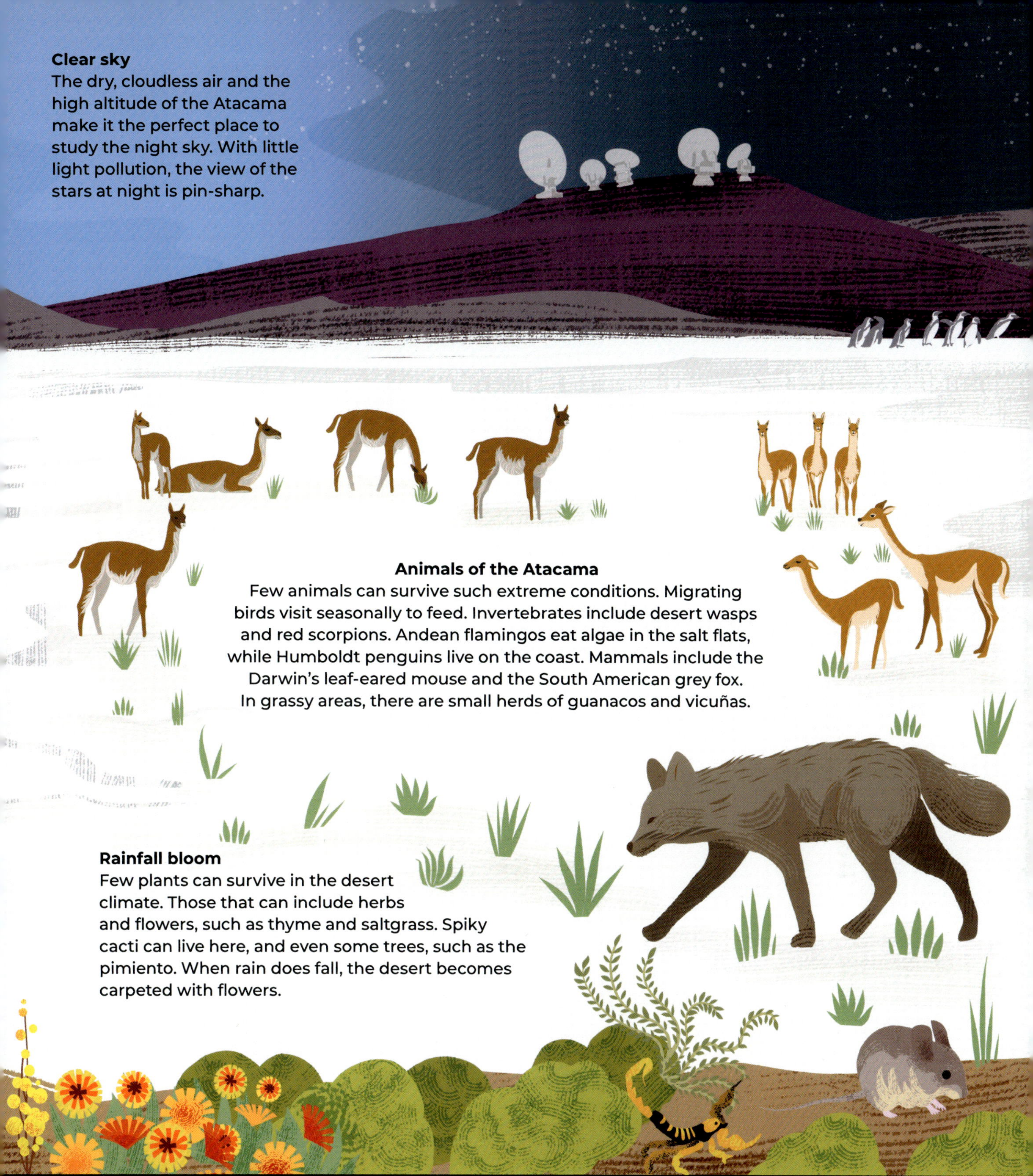

Clear sky

The dry, cloudless air and the high altitude of the Atacama make it the perfect place to study the night sky. With little light pollution, the view of the stars at night is pin-sharp.

Animals of the Atacama

Few animals can survive such extreme conditions. Migrating birds visit seasonally to feed. Invertebrates include desert wasps and red scorpions. Andean flamingos eat algae in the salt flats, while Humboldt penguins live on the coast. Mammals include the Darwin's leaf-eared mouse and the South American grey fox. In grassy areas, there are small herds of guanacos and vicuñas.

Rainfall bloom

Few plants can survive in the desert climate. Those that can include herbs and flowers, such as thyme and saltgrass. Spiky cacti can live here, and even some trees, such as the pimiento. When rain does fall, the desert becomes carpeted with flowers.

THE ATACAMA IS THE OLDEST DESERT ON EARTH, HAVING EXPERIENCED DRY CONDITIONS FOR ABOUT **150 MILLION YEARS**

TEMPERATURES DURING THE DAY IN THE ATACAMA DESERT CAN REACH 40°C, BUT DROP TO JUST 5°C AT NIGHT.

SOME PARTS OF THE DESERT ARE SO DRY THAT NO ANIMALS OR PLANTS CAN SURVIVE THERE.

Temperate rainforest

LAND OF THE GIANTS

Temperate rainforests are found in cooler regions where there is lots of rain. The Pacific coast of North America is home to one of the largest temperate rainforests on the planet.

Heavy rains
Powerful storms push plenty of rain over the temperate rainforests. Some of them receive more than 550 cm of rain every year.

Towering trees
The Pacific temperate rainforests are home to giant sequoia trees including the world's tallest tree, Hyperion, a coast redwood which is nearly 116 m tall. Another giant sequoia, General Sherman, is the world's largest with a volume of around 1,500 cubic m.

NORTH AMERICA

SPIRIT BEARS ARE A SPECIAL TYPE OF BLACK BEAR THAT HAVE WHITE FUR.

Spirit bears are unique to the Great Bear Rainforest in Canada.

Rainforest birds
Located close to coasts, these rainforests are home to seabirds, such as brown pelicans. Further inland are freshwater birds, such as ospreys and great blue herons. Birds of prey include the northern spotted owl and the bald eagle.

The forest floor
Ferns thrive on the damp forest floor, and mosses and fungi cover fallen tree trunks, which are home to hundreds of species of invertebrates.

Rainforest mammals
Mammals include northern flying squirrels, mountain beavers, deer and elk, as well as predators, such as bears, wolves, mountain lions, coyote and cougars.

MOUNTAIN LIONS, BEARS AND WOLVES ARE AMONG THE BIGGEST TEMPERATE RAINFOREST PREDATORS.

Mountain lion   Brown bear   Wolf

A MOUNTAIN LION CAN GROW TO BE OVER **1.8 M LONG.**

Mountains

TOWERING KILIMANJARO

Towering over the African grasslands, Kilimanjaro is a dormant volcano and the tallest mountain in Africa. Travelling up the sides of the mountain takes you through a number of climate zones, each with its own distinct plant and animal life.

Disappearing snow
Climate change is having an effect on Kilimanjaro. Since 1912, the snow caps at the mountain's peak have lost more than 80 per cent of their mass as the snow melts away.

Alpine desert zone
The zone between 4,000 and 5,000 m receives less than 250 mm of rain a year, making it a desert.

Arctic zone
Scree slopes, made up of loose stones, form the start of this zone, before giving way to the snow-covered peak.

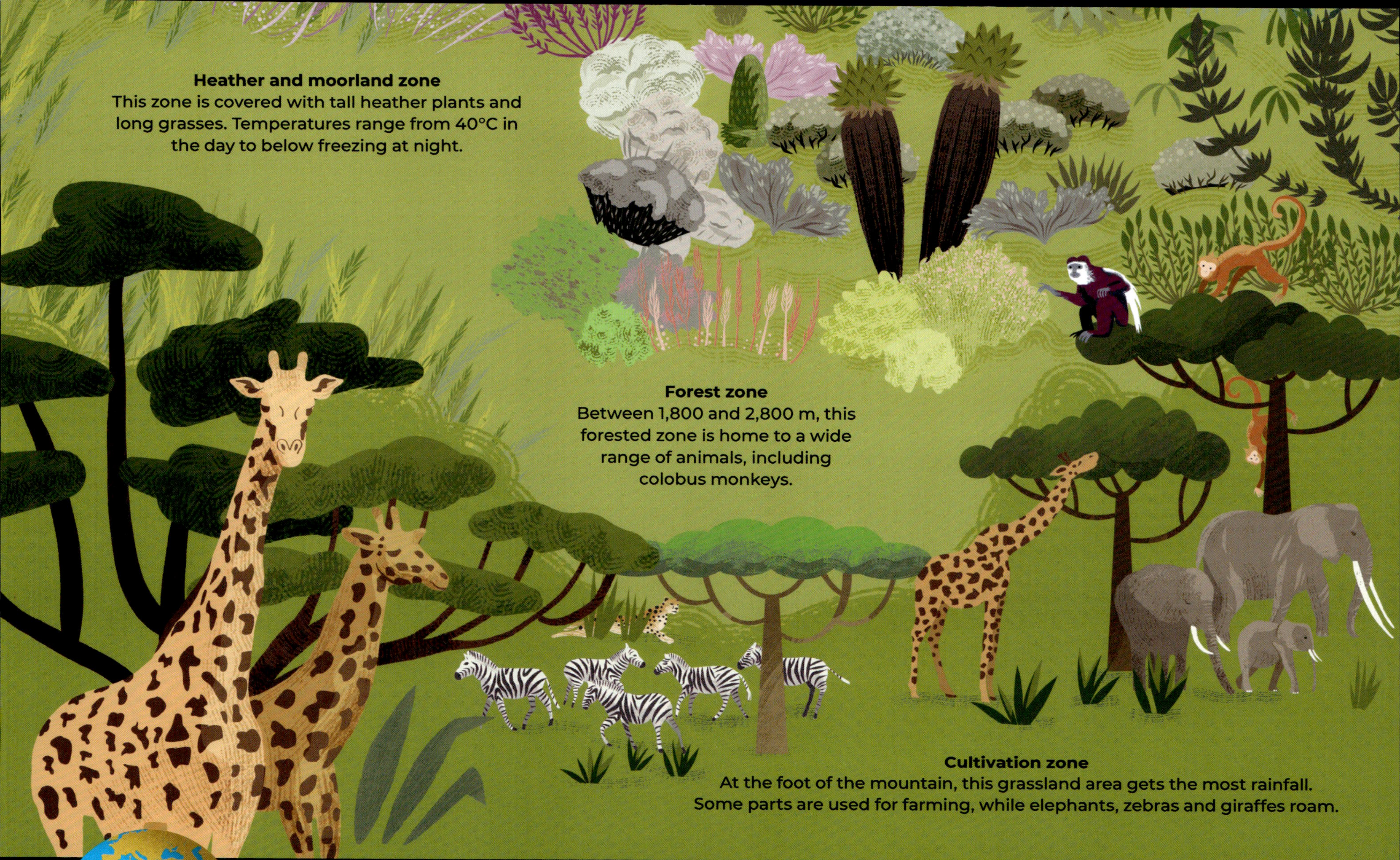

Heather and moorland zone

This zone is covered with tall heather plants and long grasses. Temperatures range from 40°C in the day to below freezing at night.

Forest zone

Between 1,800 and 2,800 m, this forested zone is home to a wide range of animals, including colobus monkeys.

Cultivation zone

At the foot of the mountain, this grassland area gets the most rainfall. Some parts are used for farming, while elephants, zebras and giraffes roam.

5,895 M

THE ALTITUDE OF THE SUMMIT, MAKING KILIMANJARO THE HIGHEST FREE-STANDING MOUNTAIN (NOT PART OF A MOUNTAIN CHAIN) IN THE WORLD.

Kilimanjaro has three volcanic cones. The last volcanic activity occurred about 200 years ago and there hasn't been a major eruption for 360,000 years.

IT IS A TYPE OF VOLCANO CALLED A STRATOVOLCANO AND STARTED TO FORM ABOUT 1 MILLION YEARS AGO.

AFRICA

Scrubland

THE AUSTRALIAN OUTBACK

Stretching across much of the interior of Australia is a vast region called the Outback. This is a mixture of desert and arid scrubland where a diverse range of plants and animals live.

Treeless land
The Nullarbor Plain is a dry, dusty part of the Outback that stretches for around 1,100 km east to west across southern and western Australia. In total, it covers about 200,000 square km.

Plants of the Nullarbor
Even though the Nullarbor Plain gets less than 200 mm of rain a year, it is home to nearly 400 plant species. Porcupine grass grows in clumps and there are a few trees, including quandong, whose sweet fruit has given it the name 'desert peach'.

AUSTRALIA

THE OUTBACK COVERS
6.5 MILLION SQUARE KM
OF THE CENTRE OF AUSTRALIA. THIS VAST AREA IS HOME TO JUST 10 PER CENT OF AUSTRALIA'S POPULATION.

The word 'Nullarbor' comes from the Latin words meaning 'no tree'.

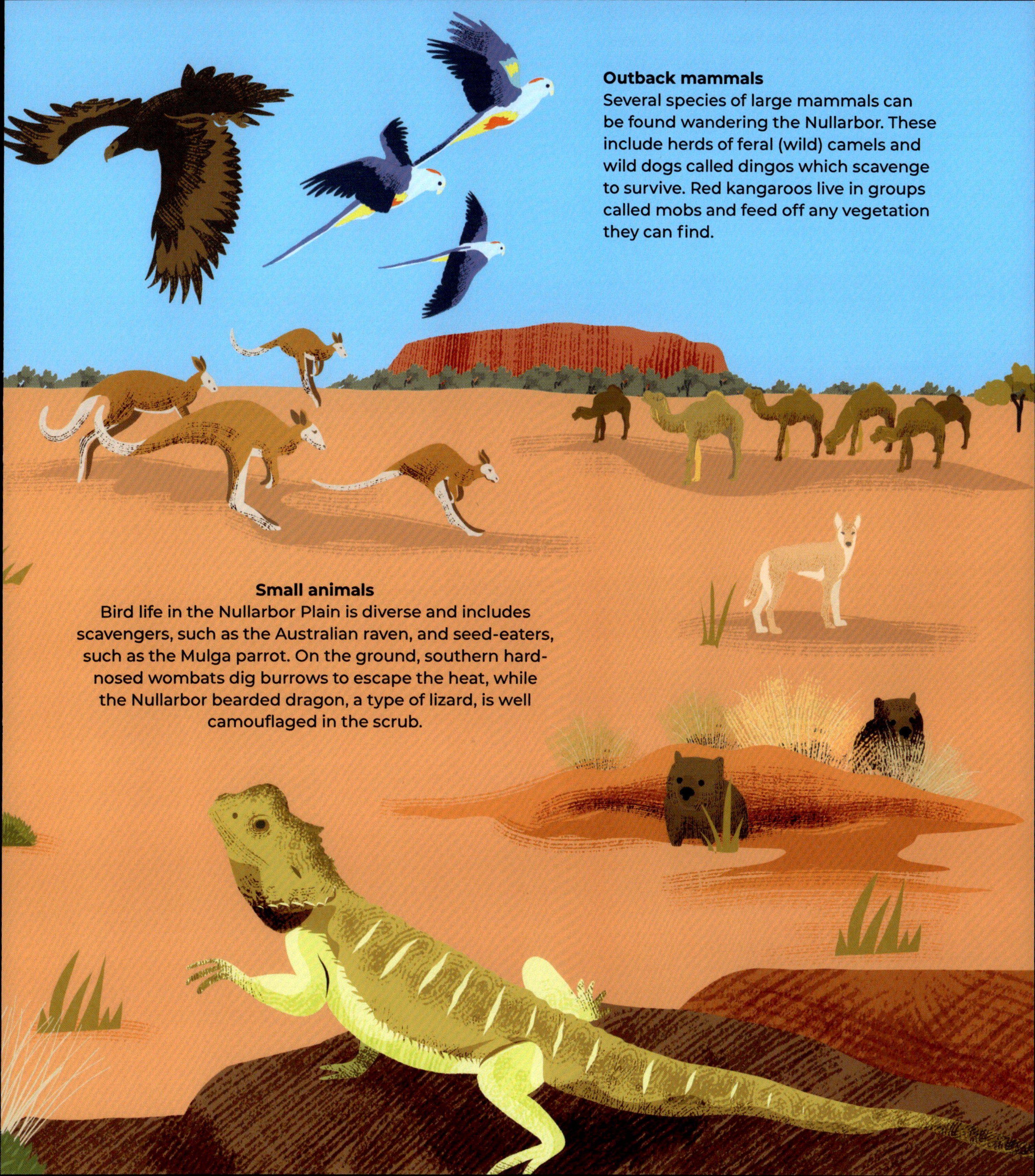

Outback mammals

Several species of large mammals can be found wandering the Nullarbor. These include herds of feral (wild) camels and wild dogs called dingos which scavenge to survive. Red kangaroos live in groups called mobs and feed off any vegetation they can find.

Small animals

Bird life in the Nullarbor Plain is diverse and includes scavengers, such as the Australian raven, and seed-eaters, such as the Mulga parrot. On the ground, southern hard-nosed wombats dig burrows to escape the heat, while the Nullarbor bearded dragon, a type of lizard, is well camouflaged in the scrub.

 AROUND 1 MILLION FERAL CAMELS LIVE IN AUSTRALIA, MAKING IT HOME TO THE WORLD'S LARGEST POPULATION OF WILD CAMELS.

RED KANGAROOS CAN BOUND ALONG AT ABOUT 55 KPH AND JUMP NEARLY 2 M VERTICALLY.

Rivers and lakes

THE MISSISSIPPI

The Mississippi River and its tributaries, including the Missouri River, form one of the world's largest river systems. It covers a huge part of North America, running through a wide range of habitats.

Source
The source of the Mississippi is Lake Itasca in Minnesota. From here, the river winds through forests, grasslands and wetlands, as well as cities, including St Louis and New Orleans.

River plants
Plants change as the river winds along its course. Tall prairie grasses and flowers give way to forests of maple and oak.

Fish
The river is home to hundreds of fish species, including carp, smallmouth bass, American eels and paddlefish.

Mississippi
Missouri
St Louis
Mississippi
New Orleans

NORTH AMERICA

Gulf of Mexico

THE MISSOURI-MISSISSIPPI RIVER NETWORK IS
5,970 KM LONG
MAKING IT THE FOURTH LONGEST RIVER NETWORK IN THE WORLD.

The Mississippi drains water from 32 US states and 2 Canadian provinces, and has a river basin covering 3.2 million square km – the fourth largest in the world.

River delta

The Mississippi empties into the Gulf of Mexico, dropping sediment to form a large delta. Many migrating birds stop off in the area to feed, including gadwalls and northern shovellers. The delta is also home to the American alligator, Louisiana black bear and green sea turtle.

Mississippi mammals

Mammal species living around the Mississippi include otters and beavers, which make their homes in the river and its banks.

Feeding birds

Many birds are attracted to the river. Mallards feed on river plants, while belted kingfishers and great blue herons catch fish. Overhead, predators, such as red-tailed hawks and bald eagles, circle.

SOME 260 SPECIES OF FISH LIVE IN THE RIVER, ABOUT A QUARTER OF ALL NORTH AMERICAN FISH SPECIES.

Carp

Paddlefish

American eel

Skipjack herring

Small-mouth bass

Tropical grassland

THE SERENGETI

This large African grassland region is home to some of the world's most amazing animals. The Serengeti is under threat, but protected areas and national parks are helping to preserve it.

Grasses and trees
The Serengeti has two types of habitat – woodland and grassland. The treeless grassland areas are dominated by long grasses, while the woodlands are dotted with trees such as acacia.

Great grazers
The Serengeti is the perfect home for grazing animals. Wildebeest eat short grasses, while zebras feed on longer ones. Small antelopes called dik-diks eat the lowest leaves, while impalas feed on the higher ones, leaving giraffes to stretch up to eat the leaves at the very top.

Growing cities
While grasslands in the protected parts remain intact, other grasslands have been swallowed up by growing towns and cities.

Lake Victoria

Serengeti National Park

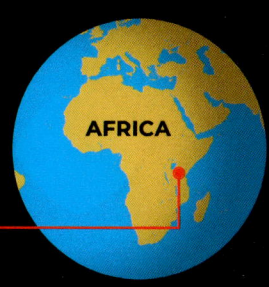

AFRICA

18,000
SQUARE KM
APPROXIMATE AREA OF THE SERENGETI

THE SERENGETI IS HOME TO:

1.3 MILLION
wildebeest

200,000
plains zebra

400,000
Thomson's gazelle

THERE ARE OVER
200
TYPES OF GRASS IN THE SERENGETI.

Bird life
Birds of the Serengeti include long-legged ostriches that can run at up to 70 kph and secretary birds that kill snakes by stamping on them with their razor-sharp claws.

CHEETAHS CAN RUN AT SPEEDS OF UP TO
120 KPH.

On the prowl
The herds of grazing animals attract predators, including lions, spotted hyenas and cheetahs. Human hunters pose an additional threat to grazing animals, as well as their predators.

The tundra
ICY CANADA

High up in the northern hemisphere lies a freezing treeless region called the tundra. In winter, the land is an icy wasteland, but during the short summer, the surface melts, flowers bloom and birds arrive to raise young.

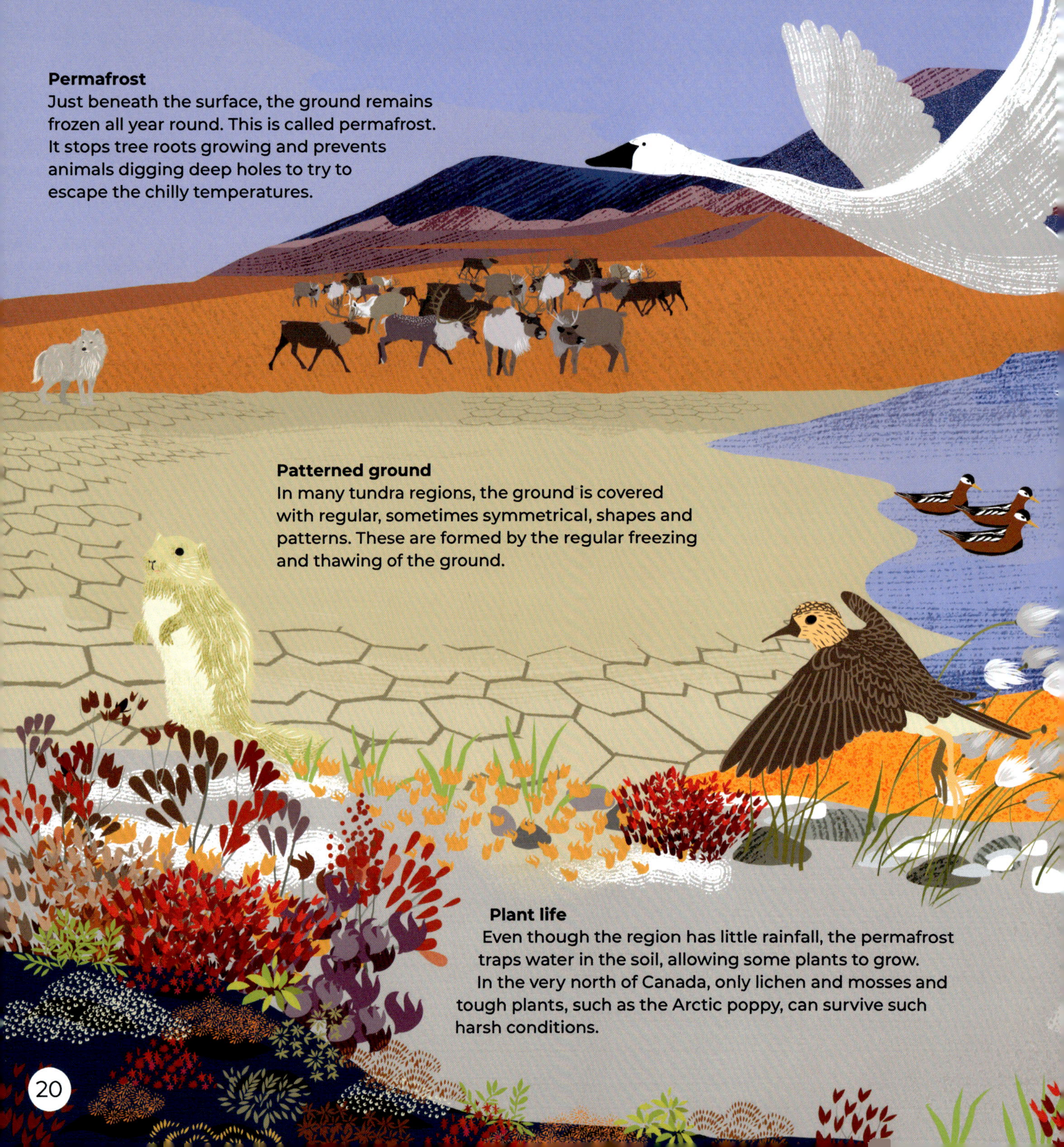

Permafrost
Just beneath the surface, the ground remains frozen all year round. This is called permafrost. It stops tree roots growing and prevents animals digging deep holes to try to escape the chilly temperatures.

Patterned ground
In many tundra regions, the ground is covered with regular, sometimes symmetrical, shapes and patterns. These are formed by the regular freezing and thawing of the ground.

Plant life
Even though the region has little rainfall, the permafrost traps water in the soil, allowing some plants to grow. In the very north of Canada, only lichen and mosses and tough plants, such as the Arctic poppy, can survive such harsh conditions.

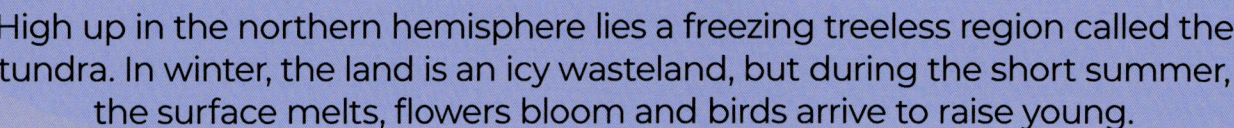

Summer visitors

Few birds stay in the tundra all year round. During the brief summer months, however, plenty of birds migrate to the region to breed as plants bloom and insects appear. Visiting birds include ducks, geese and terns.

Tundra animals

Most mammals on the tundra are migrators, such as caribou. Mammals that live here all year round include small Arctic ground squirrels and the Arctic wolf.

Changing habitats

Due to global warming, animals like the red fox are moving north into the tundra and competing with native species. Shrubs are thriving, crowding out lichens – a favourite food of the tundra's herds of caribou.

NORTH AMERICA

CLOSE TO THE ARCTIC CIRCLE, PERMAFROST CAN STRETCH MORE THAN **600 M** UNDER THE GROUND.

- -

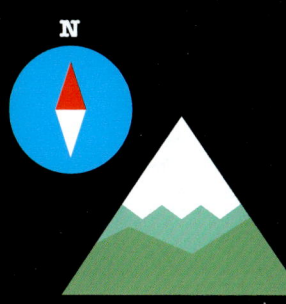

N

THERE ARE TWO TYPES OF TUNDRA – ARCTIC TUNDRA, WHICH LIES CLOSE TO THE NORTH POLE, AND ALPINE TUNDRA, WHICH IS FOUND HIGH IN MOUNTAINOUS REGIONS.

- -

The tundra gets very little precipitation (rain or snow) – just 150–250 mm per year. That's less than most of the world's deserts!

- -

A CARIBOU'S HOOVES ARE HARD IN WINTER BUT BECOME SOFTER AND SPONGIER IN SUMMER.

The Arctic
NORTHERN WASTELAND

The Arctic region is dominated by a large ocean that is covered in thick sea ice for much of the year. Surrounding this are the tops of North America, Europe and Asia, as well as hundreds of smaller islands.

Midnight Sun

Earth is tilted as it orbits the Sun. This means that in the depths of winter, the Sun does not rise at all and the Arctic experiences darkness for entire days. During summer, the Sun stays above the horizon, giving the region daylight for several days at a time.

Plant life

Arctic plant life has to cope with freezing temperatures, poor soil and permanently frozen ground. Small shrubs, mosses and algae grow here.

Bird life

The Arctic is home to water birds, such as fulmars, guillemots and eiders. There are a few land-living birds, such as the ptarmigan, as well as birds of prey, including snowy owls.

THE ARCTIC

THE ARCTIC IS HOME TO:

280 TYPES OF BIRD

450 FISH SPECIES

130 MAMMAL SPECIES

Mammals
Coastal areas are home to seals, sea lions and walruses. These attract predators, including polar bears and orcas. Reindeer and huge musk ox live inland, as well as Arctic foxes and hares.

Disappearing ice
As climate change pushes global temperatures up, more of the sea ice is melting. With less pack ice to hunt on, polar bears are forced to travel to find food wherever they can, including in towns and villages.

The name 'Arctic' comes from the Greek word 'arktos' meaning 'bear' and refers to the northern constellation Ursa Major, the Great Bear.

ARCTIC FOXES AND HARES CHANGE THE COLOUR OF THEIR COATS, FROM WHITE IN WINTER TO BROWN IN SUMMER.

Big planet journeys

All around the world, animals are moving from one place to another, making regular journeys called migrations. But why do they make these amazing and often dangerous journeys?

Drinking water
Animals often migrate to find water. Weather patterns change over the course of a year and rainfall can cease. When the rains stop, animals move to places where there is more water.

Finding food
A lack of rain also usually brings a lack of food, as plants die quickly without water. Animals have to move to places where plants are still thriving so that they have enough to eat. Warmer weather can often cause a boom in food, attracting animals from far away.

MIGRATION RECORDS

SMALLEST MIGRANT
ZOOPLANKTON
1–2 MM LONG

LARGEST MIGRANT
BLUE WHALE
UP TO 27 M LONG

Animal nursery

A place that is rich in food and water can be an ideal spot to have young and raise them, and so animals will cover enormous distances in order to breed.

Finding their way

So just how do animals find their way across the planet? Some use visible clues to navigate, such as the positions of the Sun, Moon and even the stars. Others rely on invisible pointers, including Earth's magnetic field, and even their sense of smell!

Changing times

Climate change is altering life for all living things, causing droughts, floods, food shortages and the loss of habitats. Animals are changing their migration routes in search of food and safe places to breed.

LONGEST MIGRATION
ARCTIC TERN
ABOUT 90,000 KM

HIGHEST MIGRATION
BAR-HEADED GOOSE
UP TO 10,175 M

SHORTEST BIRD MIGRATION
BLUE GROUSE
300 M

Wildebeest

CROSSING THE SERENGETI

In May and June each year, the rainy season comes to an end on the grasslands of Africa's Serengeti. Huge herds of wildebeest, zebras and other animals migrate in search of something to drink.

Disappearing water
When the rains stop, the wildebeest gather together around shrinking waterholes. The herds move when the water starts to run out. Other animals, including zebras and antelopes, join the mass movement.

Protecting the young
While a herd is on the move, the adult wildebeest surround the young, protecting them from predators, such as lions, cheetahs and hyenas.

Water hazard
Fast-flowing waters wash many wildebeest away, while hungry crocodiles wait to catch any struggling swimmers.

Mara River
Grumeti River
Serengeti National Park
TANZANIA

AFRICA

WILDEBEEST

Number of wildebeest – 1.5 million

Distance travelled – 800 km

Purpose of migration – water, food and a place to give birth

AT THE PEAK OF THE CALVING SEASON,

8,000

WILDEBEEST ARE BORN EACH DAY.

Back to the start

As the rains return in March and April, the wildebeest and their young move back to their original home. The whole journey covers about 800 km.

Calving time

From December to March, the herds are in the southern Serengeti. During this time, many of the females give birth.

On the move

The wildebeest can run at up to 80 kph and the large herds often split into smaller groups.

AROUND
200,000
ZEBRAS AND THOUSANDS OF ANTELOPES MIGRATE WITH THE WILDEBEEST.

The herds attract plenty of predators, including lions, leopards, cheetahs, crocodiles and hyenas.

Each year, about 250,000 migrating wildebeest and 30,000 zebras are killed by predators, or by thirst, hunger and exhaustion.

Humpback whales

FROM THE POLES TO THE TROPICS

Every year, humpback whales travel to the icy oceans near the North or South Poles to feed. As autumn approaches, they must migrate to warmer water to breed, before returning to the poles the following spring.

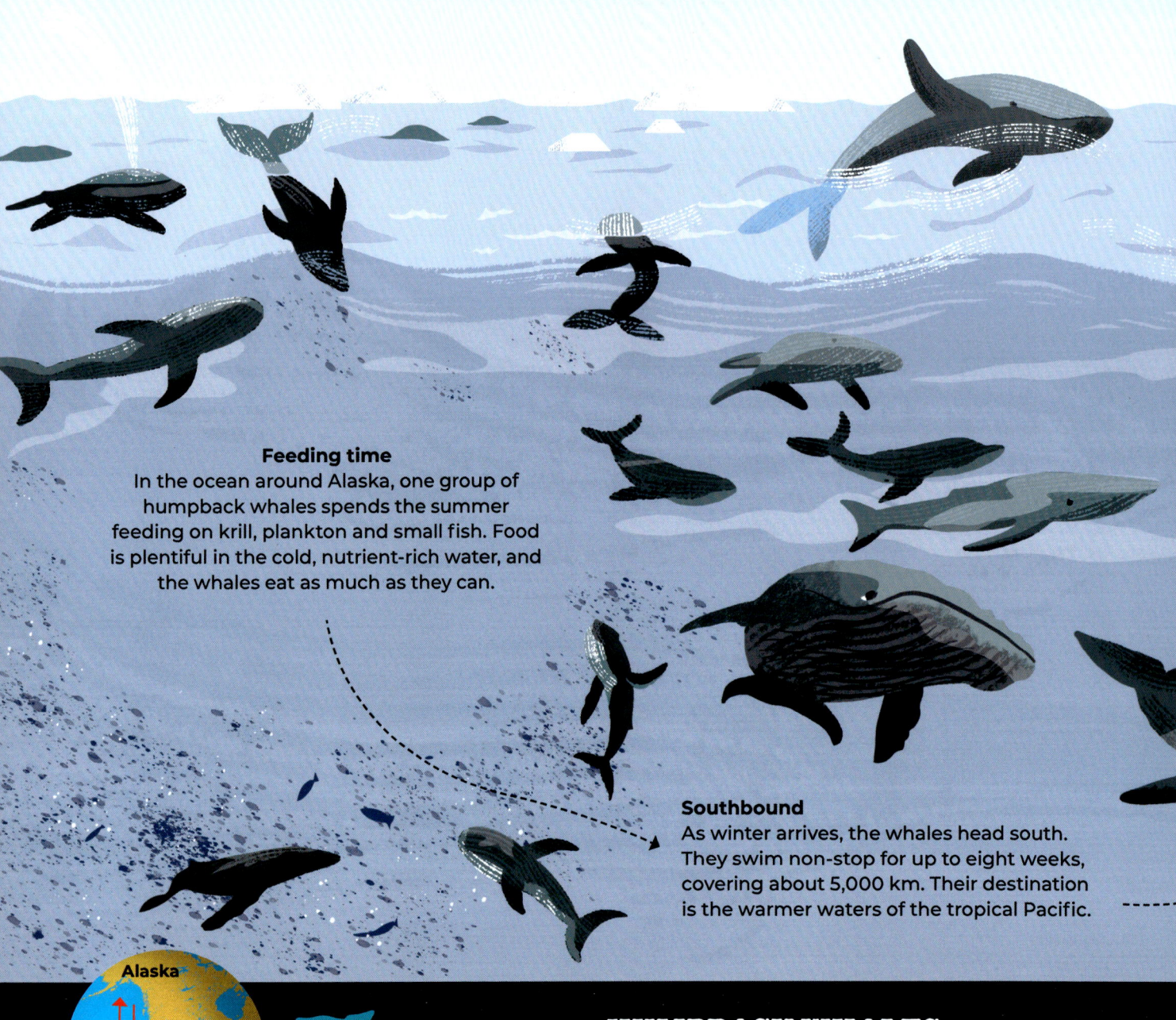

Feeding time
In the ocean around Alaska, one group of humpback whales spends the summer feeding on krill, plankton and small fish. Food is plentiful in the cold, nutrient-rich water, and the whales eat as much as they can.

Southbound
As winter arrives, the whales head south. They swim non-stop for up to eight weeks, covering about 5,000 km. Their destination is the warmer waters of the tropical Pacific.

Alaska

USA

Pacific Ocean

HUMPBACK WHALES

Number of humpback whales – around 80,000

Distance travelled – up to 25,000 km per year

Purpose of migration – to feed in polar waters and breed in tropical waters

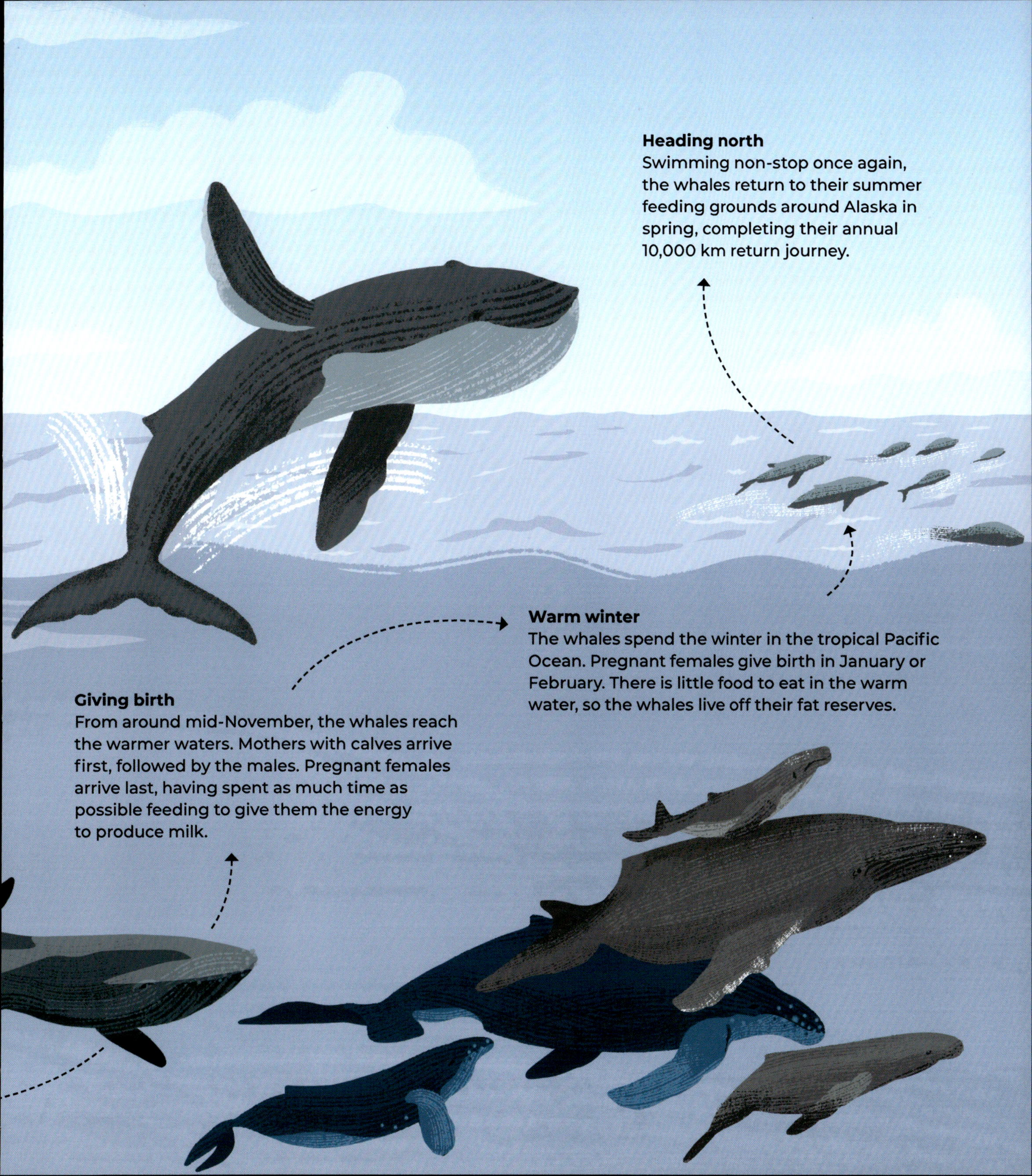

Heading north

Swimming non-stop once again, the whales return to their summer feeding grounds around Alaska in spring, completing their annual 10,000 km return journey.

Warm winter

The whales spend the winter in the tropical Pacific Ocean. Pregnant females give birth in January or February. There is little food to eat in the warm water, so the whales live off their fat reserves.

Giving birth

From around mid-November, the whales reach the warmer waters. Mothers with calves arrive first, followed by the males. Pregnant females arrive last, having spent as much time as possible feeding to give them the energy to produce milk.

DURING THE SUMMER, A HUMPBACK WHALE EATS

2.5 TONNES

OF KRILL EVERY DAY.

HUMPBACK WHALES ARE THOUGHT TO MAKE THE LONGEST MIGRATIONS OF ANY MAMMAL.

Each group of humpback whales makes a different journey. A small group in the Arabian Sea does not migrate at all.

Monarch butterflies

MEXICO TO CANADA – AND BACK!

Every year, these insects complete a remarkable journey. Over several generations, they travel from Mexico and California to Canada to breed. Then a single generation flies all the way back to avoid the chilly north!

Mexican winter
High in the mountains of central Mexico, millions of monarch butterflies gather on the branches of fir trees. Here, they huddle together to see out the winter.

New generation
In spring, the monarchs fly north to Texas, USA, where they lay eggs on milkweed plants. The eggs hatch into caterpillars.

Long migration

Near the end of summer, the monarchs reach Canada, 5,000 km north of central Mexico. These butterflies are the great-grandchildren of the insects that started the migration six months earlier!

Flying north

After a few weeks, the caterpillars pupate into butterflies. They continue to fly north to find more milkweed and lay more eggs. These eggs produce a new generation that flies further north.

Back south

The monarchs fly south to avoid winter. A single generation crosses the USA to central Mexico, where they will spend the winter before starting the next year's migration.

CANADA
USA
MEXICO

MONARCH BUTTERFLY

Number of monarch butterflies – up to 300 million

Distance travelled each day – 45 km

Purpose of migration – to avoid cold winters and breed

LIFE CYCLE

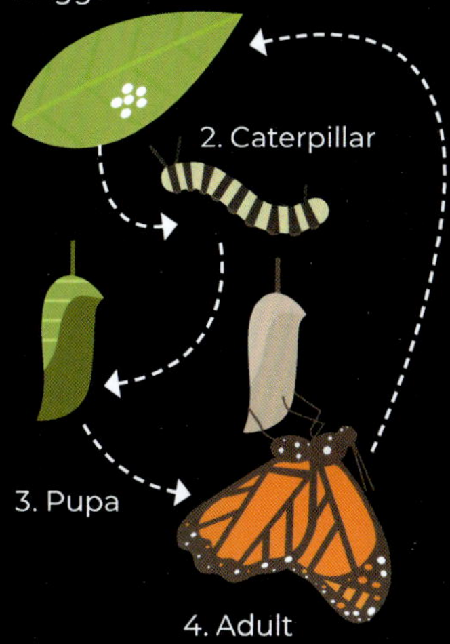

1. Eggs
2. Caterpillar
3. Pupa
4. Adult

MIGRATION MYSTERY

How do monarch butterflies find their way from Mexico to Canada and back? Scientists think they use a range of clues, such as the Sun, magnetic fields and smell. They may also use natural landmarks to guide them.

Caribou

JOURNEY ACROSS THE TUNDRA

No other land animal travels as far every year as the caribou of Alaska, USA. As spring arrives, huge herds, numbering hundreds of thousands, gather to migrate north to the tundra where they will give birth and raise their young.

Spring start
By the end of March, the caribou are at the northern edge of their winter grazing grounds. The herd then splits. Pregnant females and older caribou leave first, heading north to the coast. They are followed a few weeks later by the bulls (males) and other young caribou.

Crossing rivers
On their way, the caribou have to cross icy rivers. Their thick coats have light, hollow hairs, which keep the caribou warm and help them to stay afloat.

Alaska

USA

CARIBOU

Number of caribou – around 950,000
Distance travelled every year – 5,000 km
Purpose of migration – to calve

Climate change

With Alaska seeing warmer winters and wetter summers, the caribou's migrations are changing. Warmer weather means the caribou are staying further north for longer periods.

Return to the south

Towards the end of August, the caribou start their migration south again. They eat while moving to build up fat reserves. Males and females mate in the middle of October before the freezing winter months.

Mosquito torture

During June and July, swarms of insects, such as mosquitoes, attack the caribou. To avoid them, the caribou try to stay in windy places or keep moving.

Calving

By the end of May, the caribou have reached the calving grounds where they give birth. Here, warmer temperatures help plants such as sedges and tundra flowers to grow, providing plenty of food for caribou calves.

56 KPH
THE SPEED A CARIBOU CAN RUN AT

CARIBOU ARE KNOWN AS REINDEER IN EUROPE.

CARIBOU CALVES CAN STAND WITHIN AN HOUR OF BEING BORN.

Salmon

SWIMMING THROUGH RIVERS AND OCEANS

During their lives, Pacific salmon go through amazing changes as they migrate thousands of kilometres from rivers and streams in North America out to the Pacific Ocean, before returning to breed in the streams where they hatched.

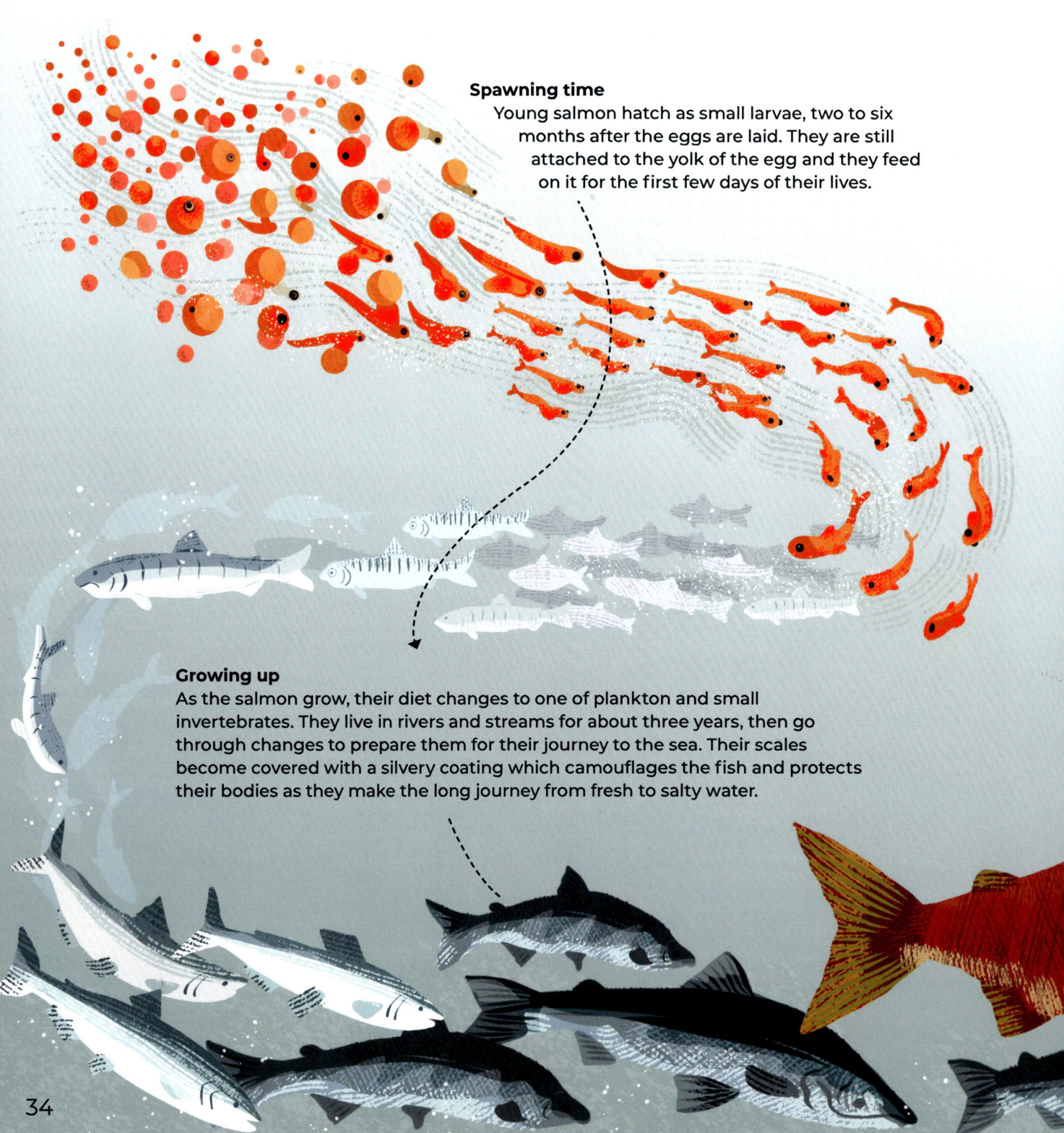

Spawning time
Young salmon hatch as small larvae, two to six months after the eggs are laid. They are still attached to the yolk of the egg and they feed on it for the first few days of their lives.

Growing up
As the salmon grow, their diet changes to one of plankton and small invertebrates. They live in rivers and streams for about three years, then go through changes to prepare them for their journey to the sea. Their scales become covered with a silvery coating which camouflages the fish and protects their bodies as they make the long journey from fresh to salty water.

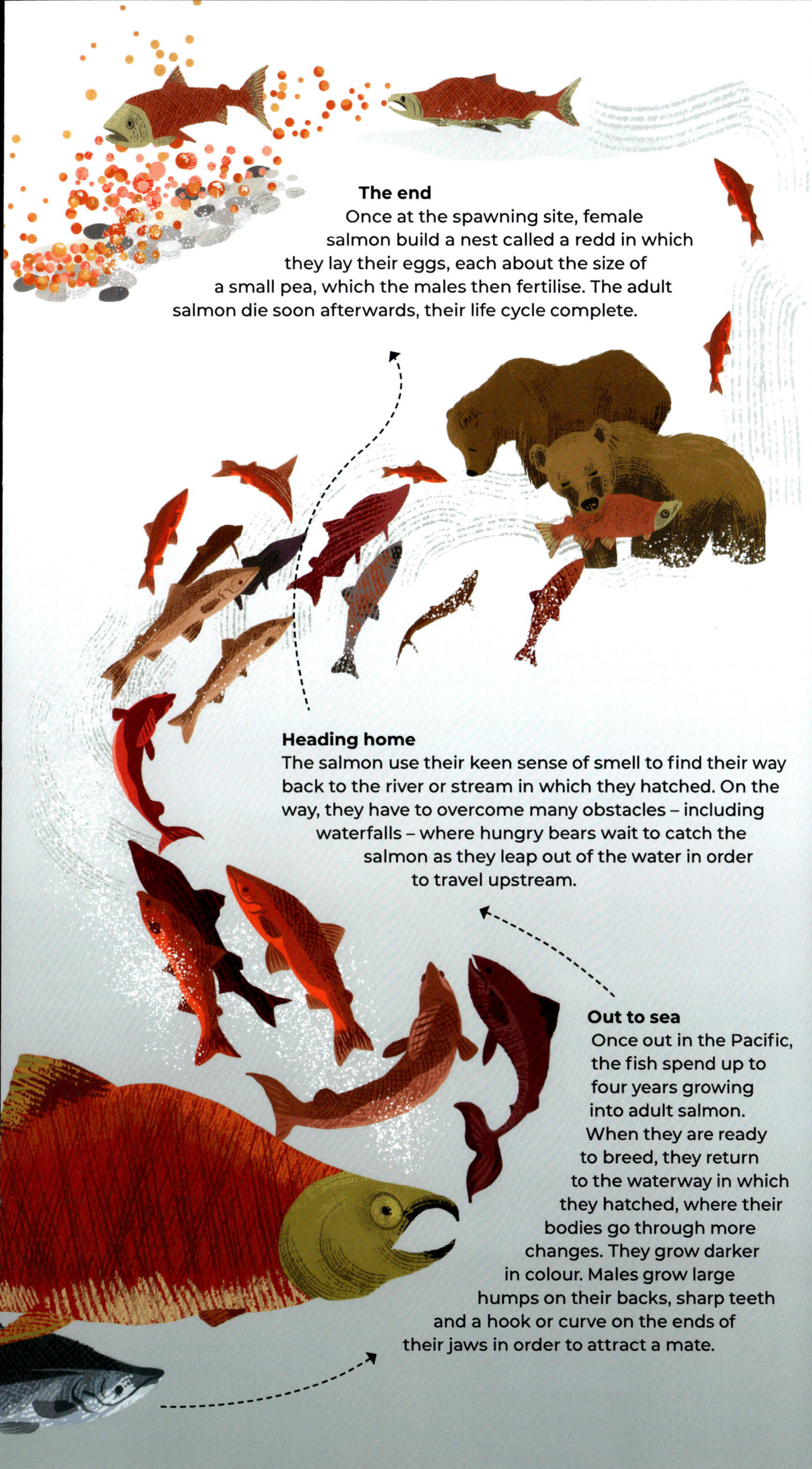

The end

Once at the spawning site, female salmon build a nest called a redd in which they lay their eggs, each about the size of a small pea, which the males then fertilise. The adult salmon die soon afterwards, their life cycle complete.

Heading home

The salmon use their keen sense of smell to find their way back to the river or stream in which they hatched. On the way, they have to overcome many obstacles – including waterfalls – where hungry bears wait to catch the salmon as they leap out of the water in order to travel upstream.

Out to sea

Once out in the Pacific, the fish spend up to four years growing into adult salmon. When they are ready to breed, they return to the waterway in which they hatched, where their bodies go through more changes. They grow darker in colour. Males grow large humps on their backs, sharp teeth and a hook or curve on the ends of their jaws in order to attract a mate.

NORTH AMERICA

Pacific Ocean

SALMON

Size of a Chinook salmon – up to 1.5 m long and 60 kg in weight

Distance travelled – 16,000 km

Purpose of migration – to feed and grow into an adult

5,000

THE NUMBER OF EGGS EACH FEMALE SALMON LAYS IN ITS REDD.

SCIENTISTS BELIEVE THAT SALMON CAN DETECT EARTH'S MAGNETIC FIELD AND USE IT TO FIND THE OPENING TO THE RIVER THEY WERE BORN IN.

SALMON HAVE BEEN SEEN JUMPING VERTICALLY ALMOST

4 M.

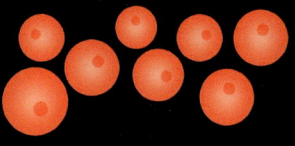

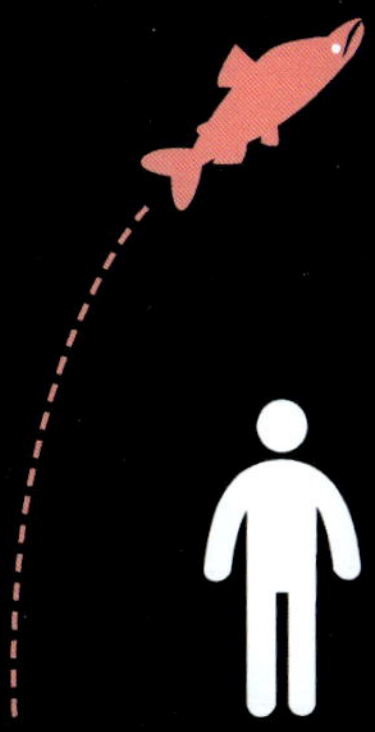

Arctic terns
LONGEST ANIMAL MIGRATION

With a migration route from the Arctic to Antarctica and back again, these birds cover up to 90,000 km every year on their hunt for food – more than twice around the planet and further than any other animal.

Breeding time
Arctic terns arrive at their northern breeding grounds in May and early June. The females lay between one and three eggs in a nest scraped into the ground. After they hatch, the adults feed their young on small fish and sea invertebrates. The young terns are ready to fly when they are a few weeks old.

Heading south
By around August, the birds are ready to fly south. The trip can take several months and many terns will wander far from the usual route. One tern from the UK was discovered 22,000 km away in Australia!

ARCTIC
AFRICA
SOUTH AMERICA
ANTARCTICA

ARCTIC TERN

Number of Arctic tern – around 1,000,000
Total distance travelled – 90,000 km
Purpose of migration – to feed and find warmer weather

30 The average age an Arctic tern can reach.

Back north again

As the southern summer comes to a close in February or March, the birds find their way back to the Arctic along the same routes they followed south. Incredibly, the birds always return to the same breeding grounds, even after a journey of tens of thousands of kilometres.

Down to Antarctica

On the way south, terns usually follow the coasts of the major continents. They then spread out over the Southern Ocean where they feed. To eat, they dip down to the water's surface to snatch up small fish.

BY TRAVELLING FROM THE ARCTIC SUMMER TO THE ANTARCTIC SUMMER, ARCTIC TERNS SEE MORE DAYLIGHT PER YEAR THAN ANY OTHER ANIMAL.

OVER ITS LIFETIME, AN ARCTIC TERN CAN FLY UP TO

2,500,000 KM.

THAT'S ALMOST THE SAME DISTANCE AS FLYING TO THE MOON AND BACK THREE TIMES!

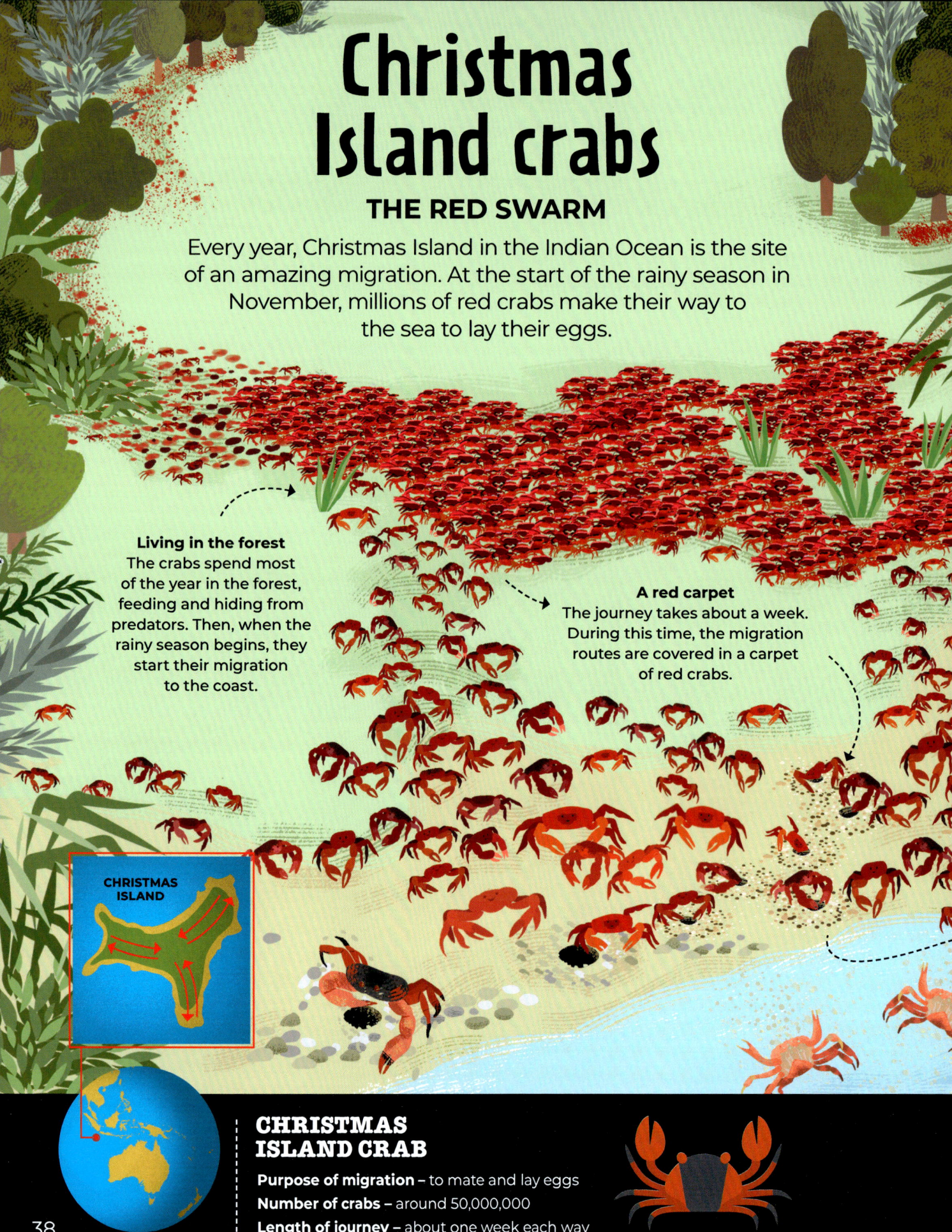

Christmas Island crabs

THE RED SWARM

Every year, Christmas Island in the Indian Ocean is the site of an amazing migration. At the start of the rainy season in November, millions of red crabs make their way to the sea to lay their eggs.

Living in the forest
The crabs spend most of the year in the forest, feeding and hiding from predators. Then, when the rainy season begins, they start their migration to the coast.

A red carpet
The journey takes about a week. During this time, the migration routes are covered in a carpet of red crabs.

CHRISTMAS ISLAND

CHRISTMAS ISLAND CRAB

Purpose of migration – to mate and lay eggs
Number of crabs – around 50,000,000
Length of journey – about one week each way

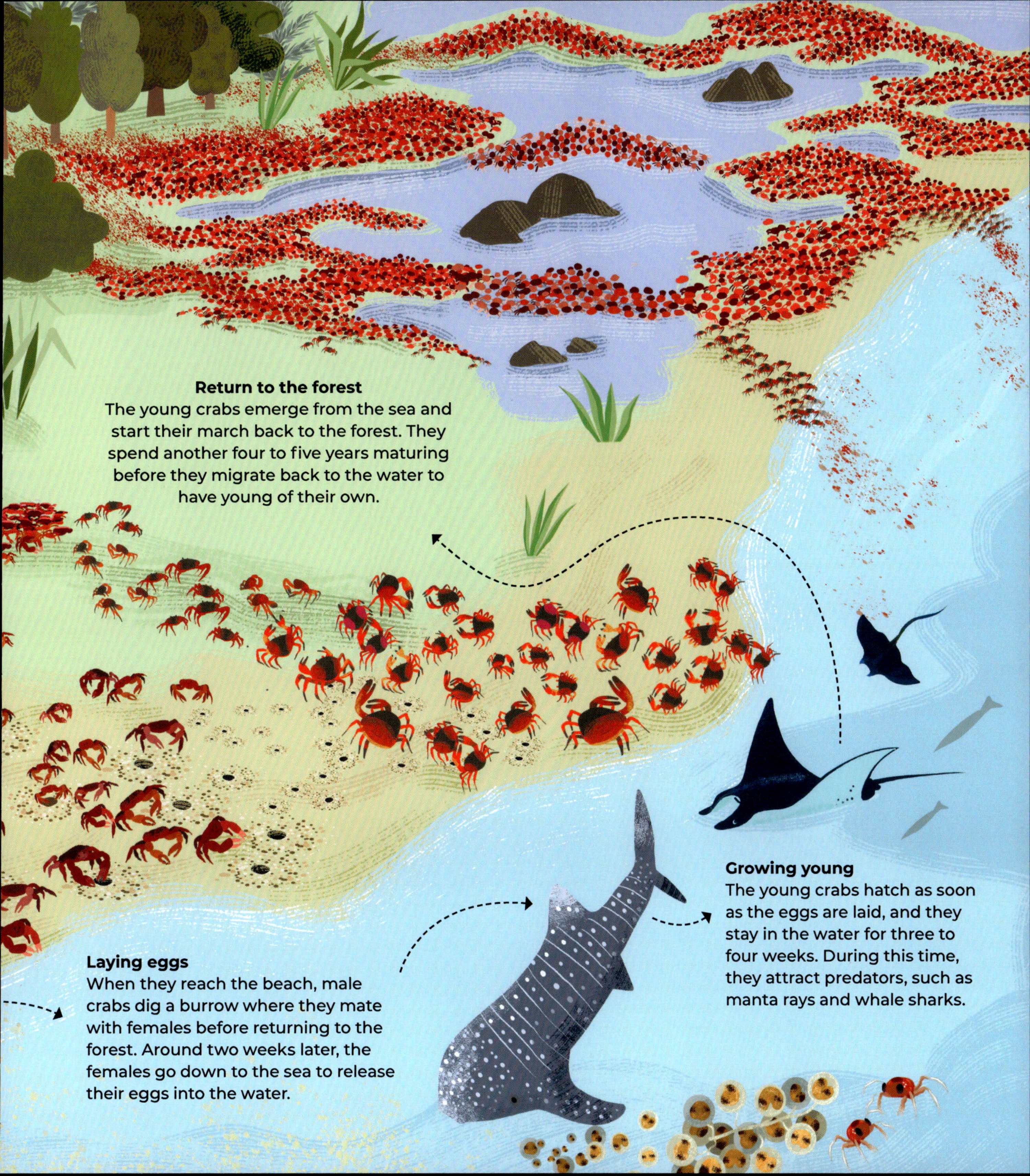

Return to the forest

The young crabs emerge from the sea and start their march back to the forest. They spend another four to five years maturing before they migrate back to the water to have young of their own.

Laying eggs

When they reach the beach, male crabs dig a burrow where they mate with females before returning to the forest. Around two weeks later, the females go down to the sea to release their eggs into the water.

Growing young

The young crabs hatch as soon as the eggs are laid, and they stay in the water for three to four weeks. During this time, they attract predators, such as manta rays and whale sharks.

ISLANDERS HAVE BUILT 'CRAB GRIDS' – SPECIAL UNDERPASSES THAT GO UNDER ROADS – AND EVEN BRIDGES SO THAT THE CRABS CAN CROSS ROADS SAFELY.

The accidental introduction of the yellow crazy ant from Australia led to a 30 per cent reduction in the red crab population. Tiny wasps were later introduced in an effort to control the ant population.

Army ants

MIGRATION OF DESTRUCTION

Swarming across the forest floor, millions of army ants carve a path of destruction to keep their colony alive. These terrifying insects raid one area before moving to another. Don't get in their way!

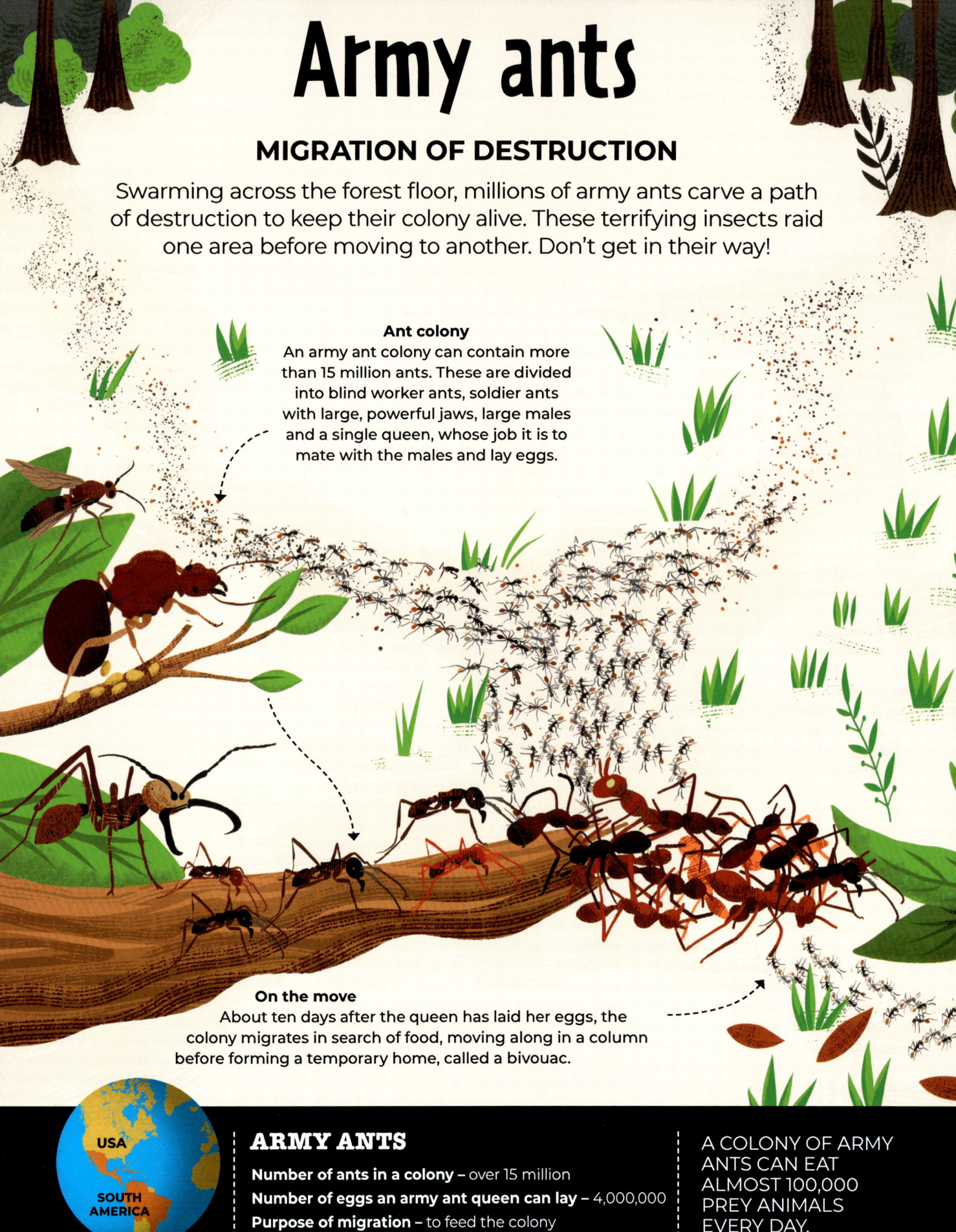

Ant colony
An army ant colony can contain more than 15 million ants. These are divided into blind worker ants, soldier ants with large, powerful jaws, large males and a single queen, whose job it is to mate with the males and lay eggs.

On the move
About ten days after the queen has laid her eggs, the colony migrates in search of food, moving along in a column before forming a temporary home, called a bivouac.

USA

SOUTH
AMERICA

ARMY ANTS

Number of ants in a colony – over 15 million
Number of eggs an army ant queen can lay – 4,000,000
Purpose of migration – to feed the colony

A COLONY OF ARMY ANTS CAN EAT ALMOST 100,000 PREY ANIMALS EVERY DAY.

Taking advantage
Waiting in the trees nearby are birds who snatch up any fleeing prey animals. Once they have exhausted the local food supply, army ants migrate to another area.

Hunting parties
Columns of workers go out to forage for food. These columns can be 20 m wide and more than 100 m long! In a single hour, an army ant raid can carry off 3,000 prey animals, from other insects to small birds and mammals.

Living camp
The bivouac is formed from millions of ants. They lock their jaws, legs and other body parts together to make a living structure. Inside are chambers for the food, larvae and eggs, and the queen. Soldier ants stand by, ready to defend the bivouac with their powerful jaws.

A QUEEN ARMY ANT CAN LIVE FOR

10–20
YEARS.

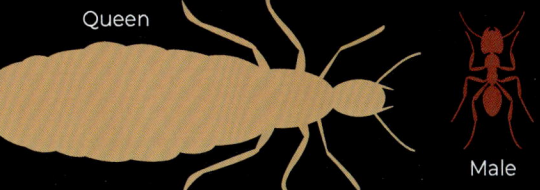

Queen

Male

As well as bivouacs, army ants build other structures using their bodies, including tunnels and bridges.

Emperor penguins

MARCH THROUGH THE ICY WORLD

While most animals migrate to warmer regions during the winter, emperor penguins march into the frozen Antarctic to raise their young. But as climate change leads to sea ice melting, we are losing the habitat that emperor penguins rely on.

Feeding time
From January to March, the adult penguins feed and mate in the Antarctic waters, which are rich in nutrients and fish during these warmer summer months.

The long march
At the end of March, the adults migrate about 150 km inland to their nesting grounds. Here, the females lay one egg each and pass them on to the males who store their eggs in incubating pouches to keep them warm.

Atlantic Ocean

ANTARCTICA

Southern Ocean

EMPEROR PENGUINS

Number of emperor penguins – over 590,000
Distance travelled – about 160 km each way
Purpose of migration – to raise their young

During the winter months, Antarctic temperatures can drop to **-60°C.**

To the sea

By December, the chicks have grown and replaced their downy feathers with waterproof ones. The parents and young then march back to the Southern Ocean to feed in the sea.

The return

The eggs start to hatch in August and the females return. Among the huge crowds, they locate their mate and their young by their calls. The males now leave to feed, while the females feed the chicks on pieces of fish they bring up from their stomachs.

Long cold winter

The female emperor penguins then migrate back across the ice, walking and sliding their way to the sea to feed. As temperatures plummet, the males huddle together in their thousands to keep themselves – and their eggs – warm.

By the end of the winter, male penguins can lose up to 40 per cent of their body weight.

EMPEROR PENGUINS ARE THE LARGEST PENGUIN SPECIES, STANDING ABOUT 1.2 M TALL.

THEY CAN DIVE TO DEPTHS OF MORE THAN 500 M AND STAY UNDERWATER FOR MORE THAN 20 MINUTES.

Big blue planet

Earth is a blue planet. Around 70 per cent of its surface is covered in water. Most is found in its five huge oceans, but there are also smaller pockets of water found in seas, lakes and rivers. Within these watery worlds is a huge range of habitats that are home to an even bigger range of living things.

Coastal habitats
Where the sea meets land can be a brutal place to live, as waves pound the shore and the tides pull the water in and out.

Close to shore
Near the shore, water can be warmer and more hospitable. These areas are teeming with coral reefs, lush seagrass meadows and thick kelp forests.

STARTING WITH THE BIGGEST, THE FIVE OCEANS ARE:

Pacific Ocean 168,723,000 square km

Atlantic Ocean 85,133,000 square km

Indian Ocean 70,560,000 square km

Southern Ocean 21,960,000 square km

Arctic Ocean 15,558,000 square km

Out at sea
Far from land, the animals of the open ocean are scattered over a vast area.

Deep dark place
Far below the surface, the oceans are dark and cold. No light reaches below a depth of 1,000 m and the weight of the water exerts crushing pressure. Even so, many living things make the deep ocean their home.

Under threat
Global warming is seeing a steady rise in sea levels as polar ice melts into the oceans and warmer waters expand, threatening low-lying habitats. Pollution also dumps huge amounts of waste into the oceans. Floating plastics form enormous islands of rubbish, or break down into tiny particles that can be eaten by fish and enter the food chain.

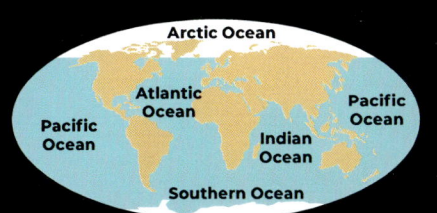

Arctic Ocean
Atlantic Ocean
Pacific Ocean
Pacific Ocean
Indian Ocean
Southern Ocean

THE OCEANS HOLD ABOUT
96.5%
OF ALL OF EARTH'S WATER.

THAT'S EQUIVALENT TO
1,338,000,000
CUBIC KM OF WATER.

Up to **12.7 million tonnes of plastic** enter the oceans every year.

River mouths

OPENING TO THE SEA

Africa's mighty River Nile flows into the Mediterranean Sea through the Nile Delta in Egypt. The delta is flat and fertile. The land here has been used for farming for thousands of years.

Forming a fan

As the Nile approaches the sea, it slows down and starts to drop a fine material called sediment. This builds up, dividing the river into smaller channels that spread out into a fan shape.

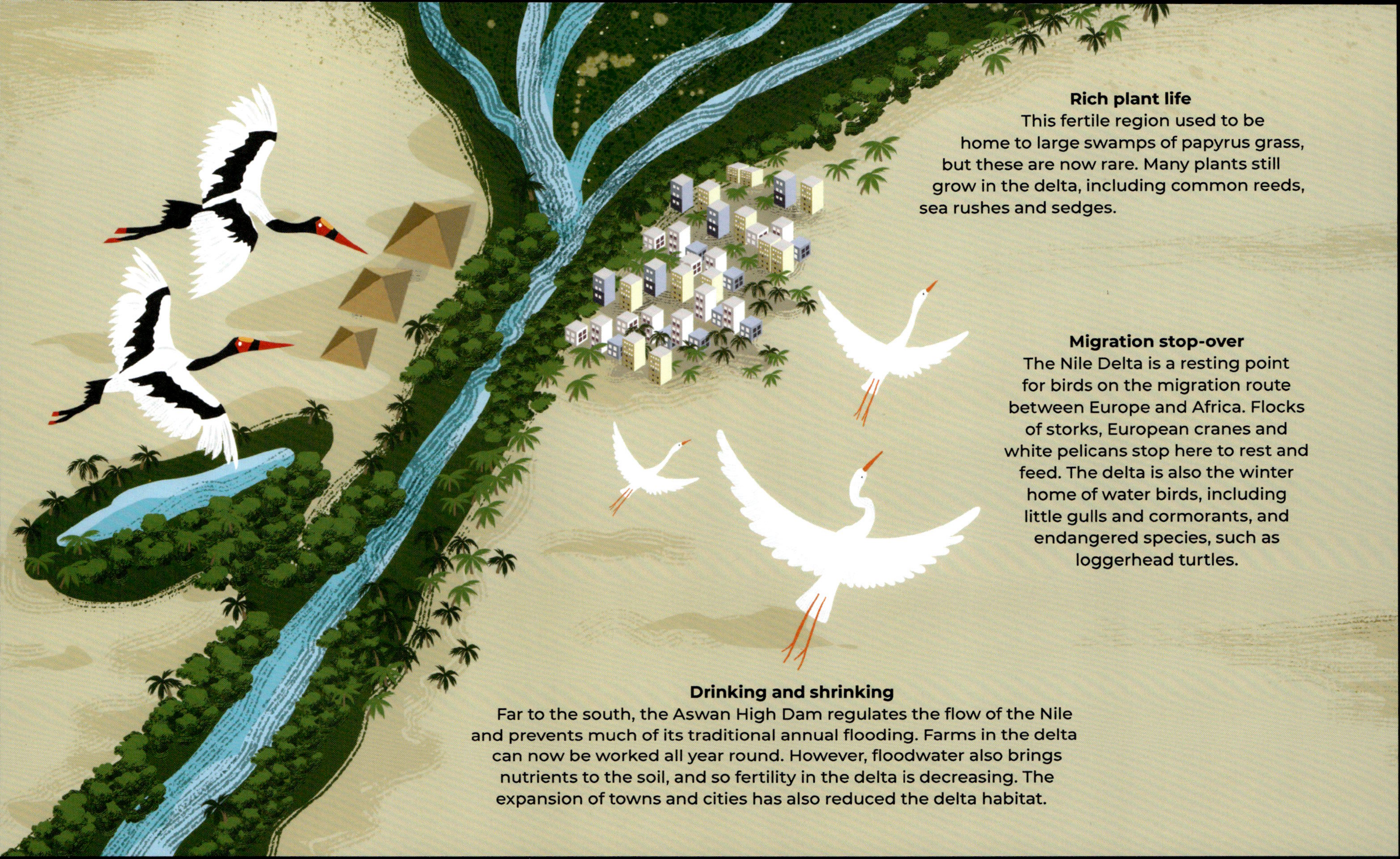

Rich plant life

This fertile region used to be home to large swamps of papyrus grass, but these are now rare. Many plants still grow in the delta, including common reeds, sea rushes and sedges.

Migration stop-over

The Nile Delta is a resting point for birds on the migration route between Europe and Africa. Flocks of storks, European cranes and white pelicans stop here to rest and feed. The delta is also the winter home of water birds, including little gulls and cormorants, and endangered species, such as loggerhead turtles.

Drinking and shrinking

Far to the south, the Aswan High Dam regulates the flow of the Nile and prevents much of its traditional annual flooding. Farms in the delta can now be worked all year round. However, floodwater also brings nutrients to the soil, and so fertility in the delta is decreasing. The expansion of towns and cities has also reduced the delta habitat.

The Nile Delta is full of wildlife, including turtles, mongooses and the Nile monitor lizard.

The Nile itself is **6,695 km long** and has a river basin that covers **3,026,000 square km.**

THE NILE DELTA IS **175 KM LONG** AND **260 KM WIDE,** COVERING AN AREA OF **26,000 SQUARE KM.**

The term **'delta'** comes from the Greek letter of the same name, which looks like a triangle.

Between the tides

A CHANGING HABITAT

When the tide is high, seawater covers the coastline, but at low tide small pools of water are left behind in the cracks and spaces between rocks. Living inside these pools is a wide range of creatures, from tiny invertebrates to fish.

Exposed rocks
The area of coast that is covered with water at high tide and exposed at low tide is called the intertidal zone. When the water retreats, anything living here has to cope with direct sunlight and pounding waves. Barnacles and limpets grip firmly to the exposed rocks, waiting for the water to return.

High tide zone
Closer to the sea, rock pools are left by the retreating tide. They are home to seaweed and an array of creatures, such as colourful anemones, mussels and crabs.

Low tide zone
This part of the shoreline remains underwater most of the time. A greater range of living things can survive here, including small fish, such as blennies, gobies and pipefish.

Rock pool hunters
Trapped in the rock pool, creatures are vulnerable to predators. Starfish crawl onto mussels, pull their shells apart and then extend their stomachs out through their mouths to eat the shellfish! Hungry gulls snatch up urchins and drop them on rocks to break open their shells to eat.

The rise and fall of the tides is caused by the gravitational pull of the Moon and the Sun.

TIDES AT THE BAY OF FUNDY IN CANADA HAVE A RANGE OF 15 M, THE GREATEST IN THE WORLD.

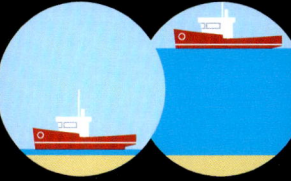

Low tide   High tide

The material that makes up limpets' teeth is the strongest known natural substance. The limpets use their teeth to scrape food such as algae off solid rock.

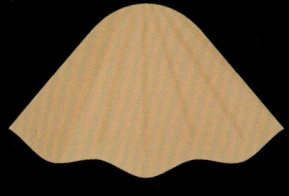

HERMIT CRABS DO NOT GROW THEIR OWN SHELL. INSTEAD, THEY USE THE EMPTY SHELLS OF OTHER ANIMALS, SUCH AS WHELKS OR SNAILS. AS THEY GROW, THEY MOVE TO LARGER SHELLS.

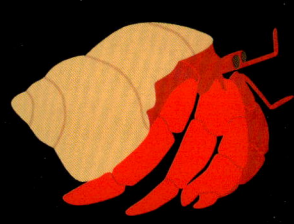

49

Mangroves

SALTY HABITATS

Mangrove forests grow along tropical coastlines where the tide floods over the land twice a day, creating a salty habitat where only a few plants can survive.

Special roots
Mangroves are trees that have adapted to living in very salty water. They stand on roots that are like stilts, keeping most of the tree above the water line, even at high tide. These wide roots have special holes in them through which the tree takes in gases, such as carbon dioxide, that they need to survive.

Snorkel roots
Some mangroves grow slim, pencil-like breathing roots that stick up out of the water. Called pneumatophores, they allow the trees to take in air from above the water.

Threatened mangroves

Mangrove forests are being cleared to make way for coastal cities and shrimp farms. Rising sea levels caused by global warming also threaten to submerge many mangrove forests completely.

Safe haven

Mangrove roots provide a safe home for invertebrates, such as brittle stars and sea urchins. They are also a great place for small fish to hide and breed. Larger fish, such as small sharks, are attracted to the mangrove forests by the rich amounts of prey.

MANGROVES LIVE IN WATER THAT'S **100 TIMES SALTIER** THAN MOST OTHER PLANTS CAN TOLERATE. THEIR ROOTS FILTER OUT THE SALT AND SOME SPECIES CAN EXCRETE IT THROUGH THEIR LEAVES.

- - - - - - - - - - - - - - - - -

35% OF THE WORLD'S MANGROVES HAVE ALREADY DISAPPEARED, AND IN MANY COUNTRIES, IT IS **AS HIGH AS 50%**.

- - - - - - - - - - - - - - - - -

Mangrove forests help to protect land during a hurricane or typhoon. They act as a buffer, absorbing much of a storm surge before it can damage the shore.

Seagrass meadows

UNDERWATER PASTURES

Just like grasslands on land, these underwater meadows provide a home both for small animals and large grazers and can be found in nearly all of the seas and oceans, from the Arctic Circle to the tropics.

Underwater plants
Seagrasses have leaves, roots and veins, just like grasses on land. They also produce flowers and seeds. Like plants on land, seagrasses make their own food using sunlight and carbon dioxide.

Smaller creatures
Seagrass provides good cover for tiny micro-organisms, like algae. These attract larger living things to feed, including crabs, sea urchins and fish. Just one hectare of seagrass can be home to 80,000 fish and 100 million small invertebrates.

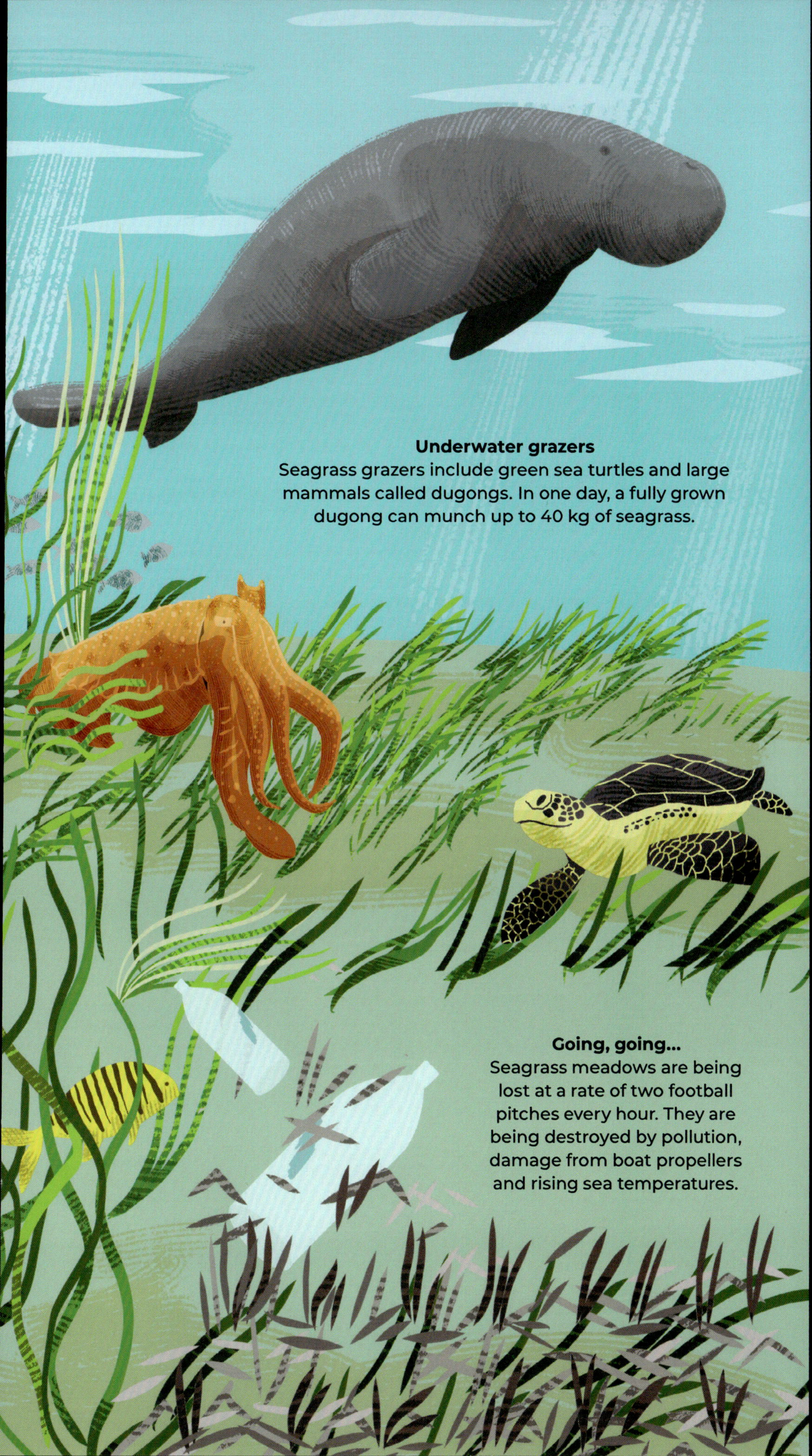

Underwater grazers

Seagrass grazers include green sea turtles and large mammals called dugongs. In one day, a fully grown dugong can munch up to 40 kg of seagrass.

Going, going...

Seagrass meadows are being lost at a rate of two football pitches every hour. They are being destroyed by pollution, damage from boat propellers and rising sea temperatures.

SEAGRASSES GROW IN SALTY OR SEMI-SALTY (BRACKISH) WATERS AROUND THE WORLD, USUALLY CLOSE TO A COASTLINE.

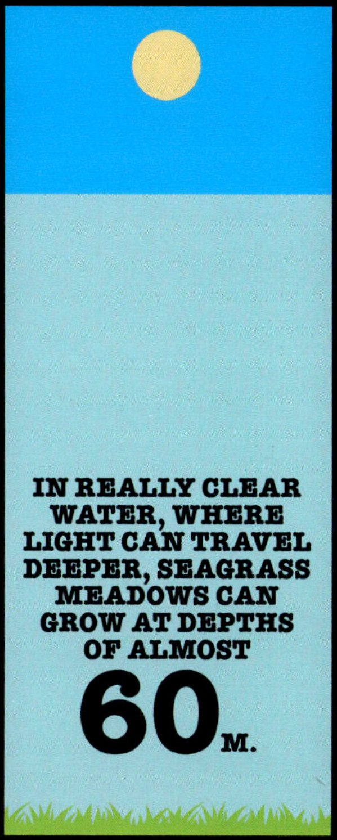

IN REALLY CLEAR WATER, WHERE LIGHT CAN TRAVEL DEEPER, SEAGRASS MEADOWS CAN GROW AT DEPTHS OF ALMOST 60 M.

72

THE NUMBER OF KNOWN SPECIES OF SEAGRASS GROWING AROUND THE WORLD.

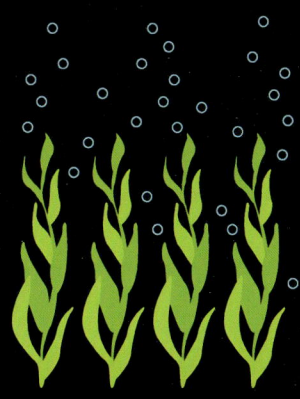

Seagrass meadows cover just 1 per cent of the sea floor, but are responsible for around 11 per cent of the carbon dioxide captured by the ocean.

Coral reefs

TEEMING WITH LIFE

Tropical coral reefs are one of the richest habitats on Earth, supporting thousands of animal and plant species. They are found in warm ocean waters close to the equator.

Tiny builders

Coral reefs are built by small creatures called polyps. These provide a home for tiny algae, which pass nutrients to the polyps in return. The algae give the coral its vibrant range of colours. Coral polyps cover themselves with a tough outer shell. Over time, these shells build up to form a reef.

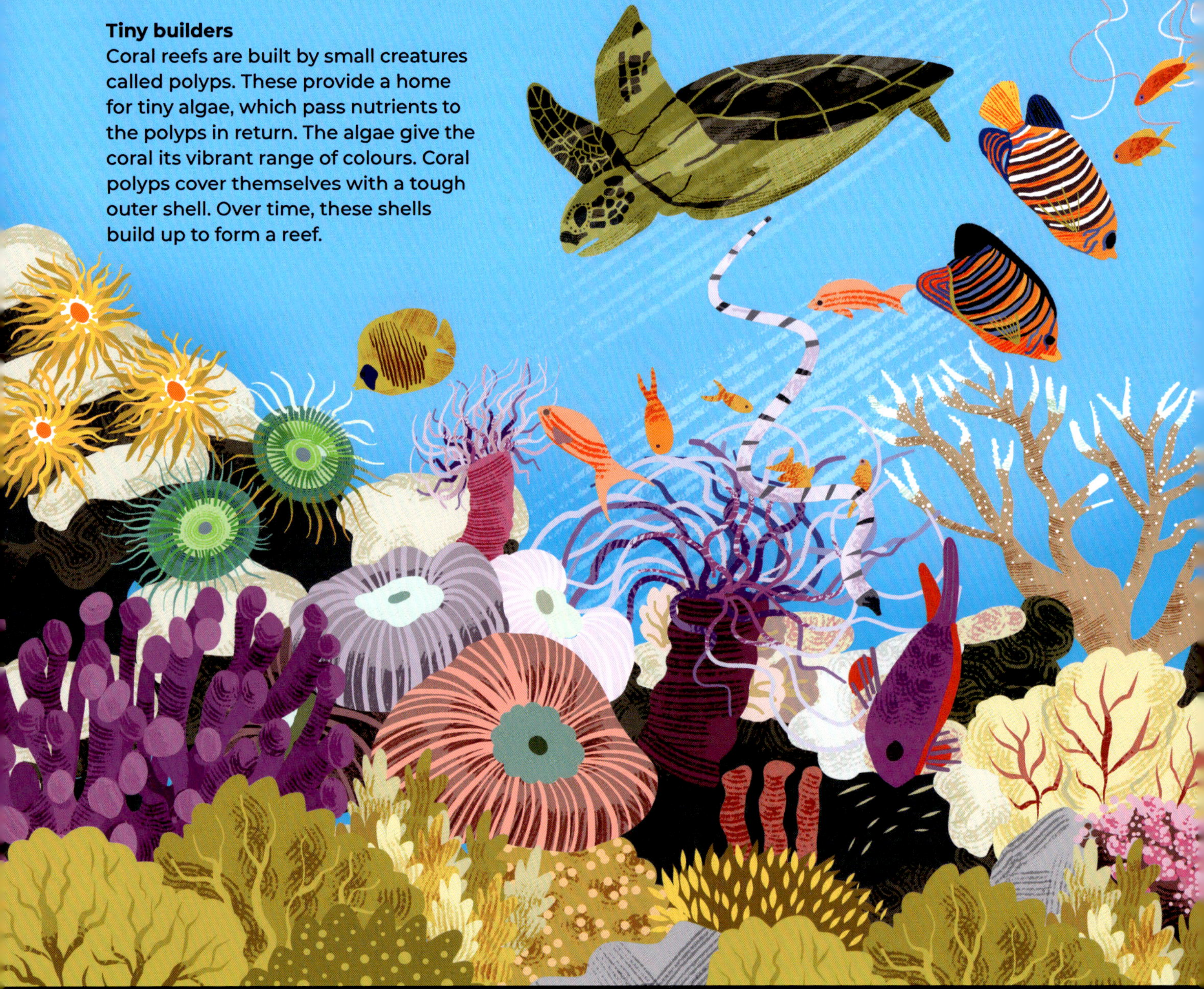

Types of reef

There are many different types of coral reef. The main ones are:

Fringing reef – a reef directly attached to the shore.

Barrier reef – a reef separated from the shore by a lagoon or water channel.

Atoll – a ring-shaped reef with a lagoon in the middle.

Wildlife paradise

With lots of nooks and crannies to hide in, and plenty of food available, coral reefs support a varied and vibrant ecosystem. Reef species range from clams and sponges to seahorses, turtles and sharks.

Under attack

Crown-of-thorns starfish can eat so much coral that their numbers are sometimes controlled by humans. Other threats to reefs include pollution, tourism and dynamite fishing. Most damaging of all, however, are rising sea temperatures, which can cause the coral to expel the algae. This is known as coral bleaching and turns healthy reefs into ghostly skeletons.

THE **GREAT BARRIER REEF** IS THE WORLD'S LARGEST CORAL REEF AND STRETCHES FOR ABOUT **2,300 KM** ALONG AUSTRALIA'S NORTHEAST COAST.

Great Barrier Reef

AUSTRALIA

Even though coral reefs make up just **0.1 per cent** of the oceans' area, they are home to about **25 per cent** of all ocean species.

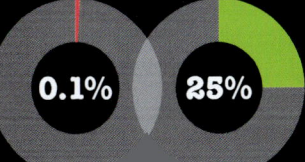

0.1%   25%

THERE ARE MORE THAN 25,000 SPECIES OF CORAL.

Open ocean

ENORMOUS ECOSYSTEM

Great expanses of the world's seas and oceans lie far from land. Although this ocean realm is a huge habitat, it is largely empty. But the presence of a food source, such as a shoal of fish, can attract many visitors.

Sardine run
In the seas off southern Africa between May and July, billions of sardines swim north in an annual migration called the sardine run. This mass migration attracts hungry predators such as dolphins and whales.

Bait ball
Dolphins herd the sardines into tight globes called bait balls. Keeping them close to the surface, the dolphins start picking off fish around the edges.

MORE THAN
50%
OF EARTH'S SURFACE IS COVERED BY OCEAN THAT IS MORE THAN **3.2 KM DEEP.**

THE **GREAT PACIFIC GARBAGE PATCH** IS THE LARGEST ACCUMULATION OF PLASTIC POLLUTION IN THE OPEN OCEAN, STRETCHING OVER **1.6 MILLION SQUARE KM**.

Attack from above

Above the surface, seabirds, such as gannets and cormorants, dive bomb the bait ball, plunging into the water to snatch up whatever fish they can.

Smash and grab

Other predators, including tuna and sharks, take advantage of all this prey, charging in to gobble up fish.

Plastic pollution

The open ocean is still affected by humans. Huge, swirling ocean currents sweep up waste from near the shore, creating floating islands of rubbish that can kill ocean wildlife that swallow it or become entangled in discarded plastic and nets.

THE OPEN OCEAN IS KNOWN AS THE PELAGIC ZONE. IT INCLUDES ALL THE OCEANS, EXCEPT FOR COASTAL WATERS AND THE SEA FLOOR.

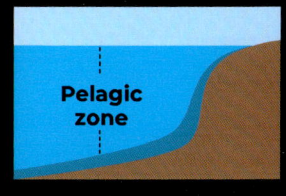

Pelagic zone

ONLY
10%
OF OCEAN SPECIES LIVE IN THIS ENORMOUS HABITAT.

Ocean deep

DARK AND CRUSHING

Deep below the surface, the water is inky black and mostly empty of life. Anything that does live down here has to deal with enormous water pressure and living in the constant dark.

Crushing pressures

At the deepest part of the ocean, pressures are about 1,100 times greater than at the surface. Deep-sea fish contain lots of water, which cannot be squashed, so they keep their shape under extreme pressure.

Lack of light

Very little sunlight reaches 200 m under the water's surface, and the water below 1,000 m is completely black. Animals that live at these depths have adapted to the dark. Some have huge eyes or eyes on stalks to collect what little light there is.

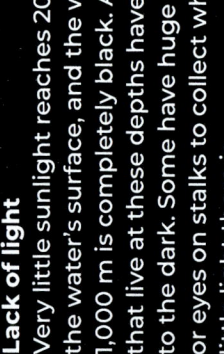

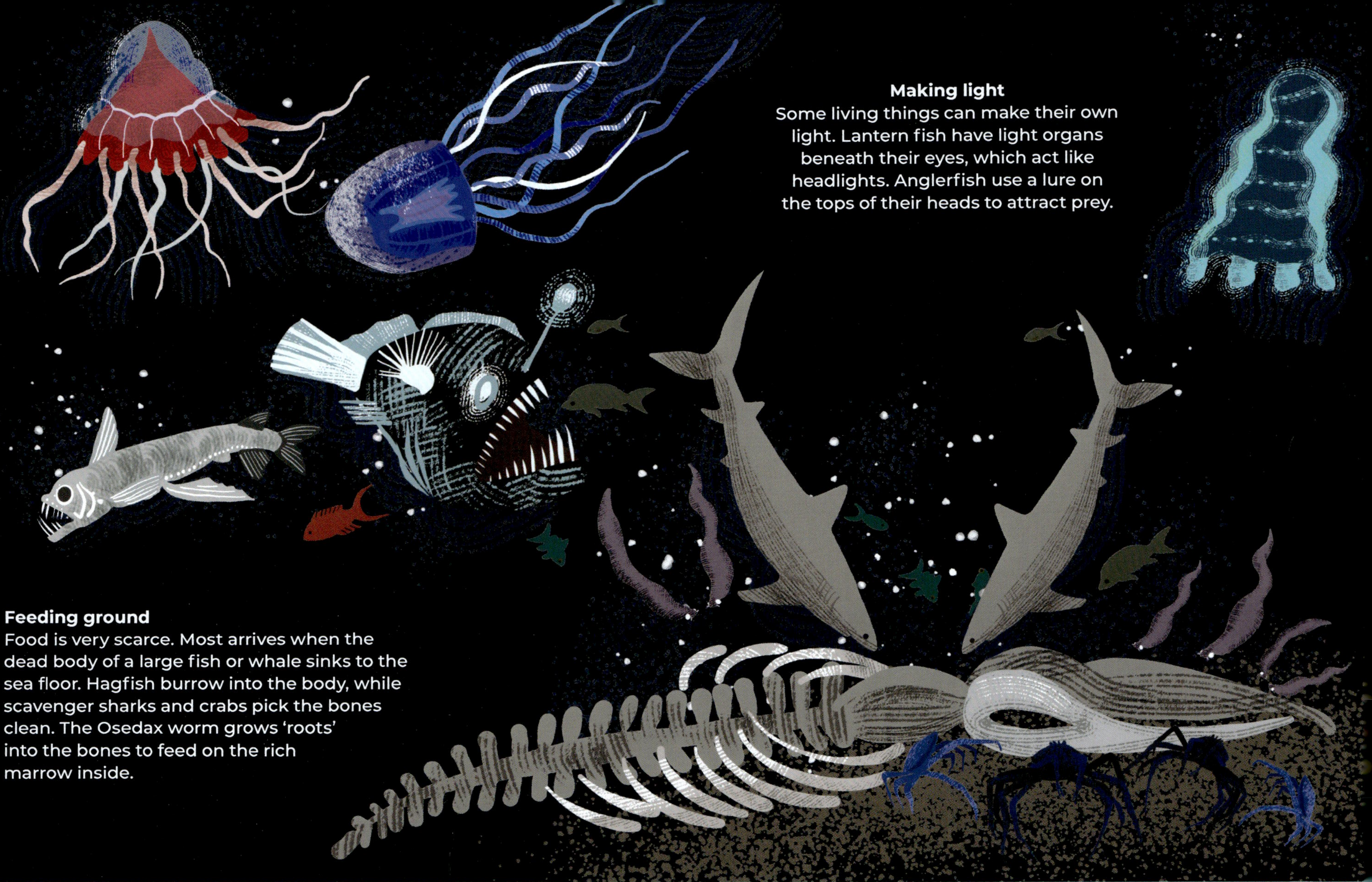

Making light

Some living things can make their own light. Lantern fish have light organs beneath their eyes, which act like headlights. Anglerfish use a lure on the tops of their heads to attract prey.

Feeding ground

Food is very scarce. Most arrives when the dead body of a large fish or whale sinks to the sea floor. Hagfish burrow into the body, while scavenger sharks and crabs pick the bones clean. The Osedax worm grows 'roots' into the bones to feed on the rich marrow inside.

When threatened, deep-sea green bomber worms release small glowing bombs to confuse predators.

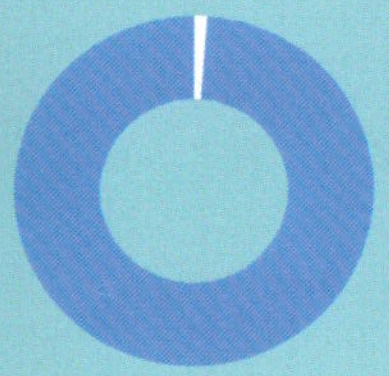

The Mariana Trench in the Pacific is the deepest part of the ocean. It lies

10,994 m

below sea-level.

99%

OF THE OCEAN FLOOR IS UNEXPLORED.

Many creatures that live in the deepest parts of the ocean are actually **TRANSPARENT.**

Southern Ocean
OCEAN FOOD CHAIN

Surrounding the Antarctic, the Southern Ocean is the windiest and coldest ocean on the planet. Despite the extreme conditions, the ocean blooms into life every summer.

Tiny treats
As summer approaches and sea ice melts, tiny micro-organisms, called phytoplankton, flourish. These become food for larger zooplankton, such as krill. Together, these tiny living things are food for fish, seabirds and even whales.

Big eaters
Fish living in the Southern Ocean include eelpouts and snailfish, but the truly big eaters are the baleen whales that migrate to the region to feed on krill. Among these are the biggest animals on Earth – blue whales. A blue whale can eat more than 3,500 kg of krill in one day!

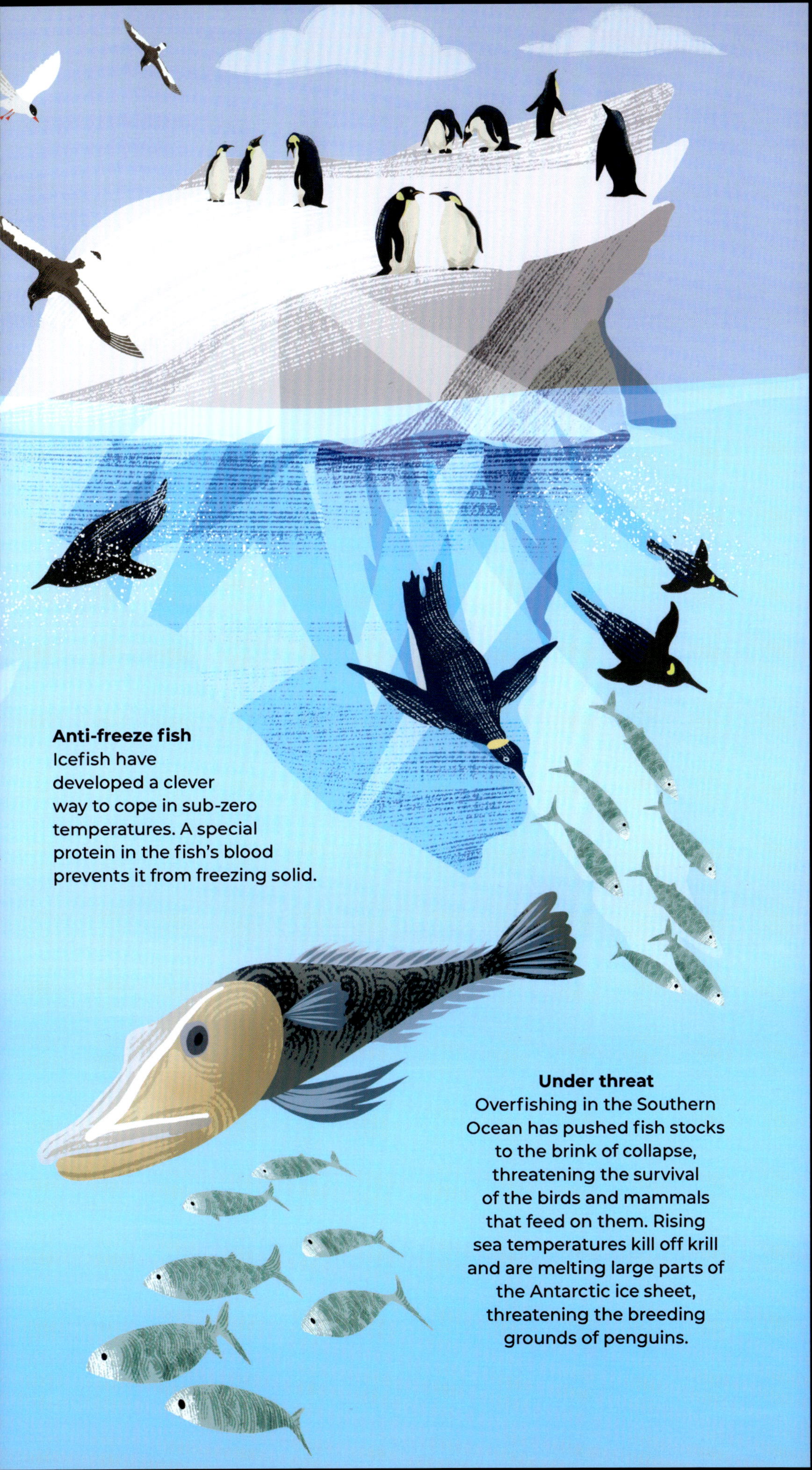

It is also the world's newest ocean. It was formed about **30 million years ago** when **Antarctica** and **South America** moved apart, and its boundaries were only proposed in 2000.

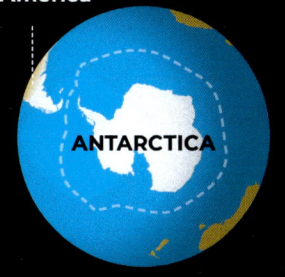

South America

ANTARCTICA

ABOUT 100 MILLION BIRDS BREED IN ANTARCTICA EACH YEAR, AND PENGUINS MAKE UP ABOUT 90 PER CENT OF THESE.

The lowest air temperature recorded in Antarctica is
-89.2°C.

Anti-freeze fish
Icefish have developed a clever way to cope in sub-zero temperatures. A special protein in the fish's blood prevents it from freezing solid.

Under threat
Overfishing in the Southern Ocean has pushed fish stocks to the brink of collapse, threatening the survival of the birds and mammals that feed on them. Rising sea temperatures kill off krill and are melting large parts of the Antarctic ice sheet, threatening the breeding grounds of penguins.

Glossary

Adapt to change over time in order to survive in a habitat

Arid very dry land, where few plants can grow

Camouflage a pattern or colouring that allows plants and animals to blend in with their environment

Canopy the layer of branches and leaves formed by trees in a forest

Climate change shifts in Earth's climate, including the rising temperatures of global warming

Delta the area where a river meets another body of water, for example the sea

Desert an area that receives less than 250 mm of precipitation (rain, sleet or snow) a year

Ecosystem a community of plants and animals that live and interact within a habitat

Endangered a species that is at risk of extinction due to environmental changes or human behaviour

Equator an imaginary line that runs around the middle of Earth, dividing it into the northern and southern hemispheres

Evaporate when a liquid, such as water, is warmed and changes into a gas

Extinction when a species of living thing dies out

Fertile land that is good for growing crops on

Fertilise when the egg from a female animal joins with the sperm from a male animal

Food chain a sequence that shows how species rely on each other for food, with smaller plants and animals being eaten by larger animals

Fungi types of living things that include tiny organisms such as mould and mushrooms

Generation a group of living things that are born at around the same time

Habitat the place where an animal or plant naturally lives

Invertebrate a type of animal that does not have a spine

FURTHER INFORMATION

WEBSITES
www.nhm.ac.uk
www.natgeokids.com
bbcearth.com/bbc-earth-kids
www.oceanservice.noaa.gov/kids/

MUSEUMS
Natural History Museum
Cromwell Road, South Kensington,
London SW7 5BD

Larva a stage in the life cycle of many invertebrates. A caterpillar is the larva of a butterfly, for example

Life cycle all the growth and changes that an animal or plant goes through from its birth until its death

Mammal a warm-blooded animal that has a backbone and grows hair at some stage of its life cycle

Micro-organisms tiny living things that include many types of bacteria, fungi and algae

Migration the seasonal movement of animals from one region to another, often to feed or breed

Minerals natural substances found in Earth's rocks, sand and soil

Nutrients natural substances that help animals and plants to grow

Poles the regions that lie at the top and bottom of Earth. The region around the North Pole is called the Arctic, while the region around the South Pole is called Antarctica

Pollution waste substances and materials, like plastic, that contaminate the natural environment

Predator an animal that hunts other living things

Prey an animal that is hunted by another for food

Rainforest a type of forest that receives a lot of rain

Scavenge to feed off scraps and waste

Species a group of living things that share characteristics and can reproduce

Temperate the regions of the world that lie between the tropics and the polar regions

Tributary a smaller river or stream that joins with a bigger one

Tropical the regions of the world that lie on either side of the equator

Tundra areas found high up mountains or close to the poles where no trees grow and the ground is frozen all year round

 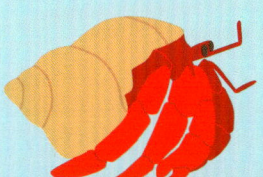   

BOOKS
Atlas of Amazing Migrations by Matt Sewell and Megan Lee (Pavilion, 2021)
A celebration of Earth's most extreme journeys.
Blue Worlds series by Anita Ganeri and Josy Bloggs (Wayland, 2022)
This series explores all aspects of the world's oceans, including habitats, species and the threats they face.
Our Planet by Matt Whyman and Richard Jones (HarperCollins, 2019)
A visual exploration of the natural world.

Index

First published in Great Britain
in 2023 by Hodder & Stoughton
Copyright © Hodder & Stoughton Limited, 2023
All rights reserved.
The material in this book has previously appeared
in the *Big Picture* series.

Series editor: Julia Bird
Produced by Tall Tree Ltd
Artist: Josy Bloggs
Printed in the UK
HB ISBN: 978 1 4451 8636 8
PB ISBN: 978 1 4451 8637 5

An Hachette UK Company
www.hachette.co.uk
www.hachettechildrens.co.uk

Franklin Watts
An imprint of Hachette Children's Group
Part of Hodder and Stoughton
Carmelite House
50 Victoria Embankment
London EC4Y 0DZ

The authorised representative in the EEA is
Hachette Ireland
8 Castlecourt Centre
Dublin 15, D15 XTP3, Ireland
email: info@hbgi.ie

The websites (URLs) included in this book were valid at
the time of going to press. However, it is possible that
the contents or addresses may have changed since the
publication of this book. No responsibility for any such
changes can be accepted by either the author or the
Publisher.

FSC MIX Paper | Supporting responsible forestry
www.fsc.org FSC® C104740